Risk, Technology, and Moral Emotions

Risks arising from technologies raise important ethical issues. Although technologies such as nanotechnology, biotechnology, ICT, and nuclear energy can improve human well-being, they may also convey risks for our well-being due to, for example, abuse, unintended side effects, accidents, and pollution. As a consequence, technologies can trigger emotions, including fear and indignation, which often leads to conflicts between stakeholders. How should we deal with such emotions in decision making about risky technologies?

This book offers a new philosophical theory of risk-related emotions, arguing why and how *moral* emotions should play an important role in decisions surrounding risky technologies. Emotions are usually met with suspicion in debates about risky technologies because they are seen as contrary to rational decision making. However, Roeser argues that moral emotions can play an important role in judging ethical aspects of technological risks, such as justice, fairness, and autonomy. This book provides a novel theoretical approach while at the same time offering concrete recommendations for decision making about risky technologies. It will be of interest to those working in different areas of philosophy—such as risk ethics, environmental ethics, and bioethics, decision theory, philosophy of science, and philosophy of technology—as well as scholars in the fields of psychology, sociology, public policy, and science and technology studies.

Sabine Roeser is Professor of Ethics in the Philosophy Department at Delft University of Technology, The Netherlands

Routledge Studies in Ethics and Moral Theory

For a full list of titles in this series, please visit www.routledge.com

30 **Environmental Skill**
Motivation, Knowledge, and the Possibility of a Non-Romantic Environmental Ethics
Mark Coeckelbergh

31 **Developing Moral Sensitivity**
Edited by Deborah S. Mower, Phyllis Vandenberg, and Wade L. Robison

32 **Duties Regarding Nature**
A Kantian Environmental Ethic
Toby Svoboda

33 **The Limits of Moral Obligation**
Moral Demandingess and 'Ought Implies Can'
Edited by Marcel van Ackeren and Michael Kühler

34 **The Intrinsic Value of Endangered Species**
Ian A. Smith

35 **Ethics and Social Survival**
Milton Fisk

36 **Love, Reason and Morality**
Edited by Esther Engels Kroeker and Katrien Scaubroeck

37 **Virtue's Reasons**
New Essays on Virtue, Character, and Reasons
Edited by Noell Birondo and S. Stewart Braun

38 **In Defense of Moral Luck**
Why Luck Often Affects Praiseworthiness and Blameworthiness
Robert J. Hartman

39 **Risk, Technology, and Moral Emotions**
Sabine Roeser

Risk, Technology, and Moral Emotions

Sabine Roeser

Routledge
Taylor & Francis Group
NEW YORK AND LONDON

First published 2018
by Routledge
711 Third Avenue, New York, NY 10017

and by Routledge
2 Park Square, Milton Park, Abingdon, Oxon OX14 4RN

Routledge is an imprint of the Taylor & Francis Group, an informa business

Library of Congress Cataloging-in-Publication Data
Names: Roeser, Sabine, author.
Title: Risk, technology, and moral emotions / by Sabine Roeser.
Description: 1 [edition]. | New York : Routledge, 2017. | Series: Routledge studies in ethics and moral theory ; 39 | Includes bibliographical references and index.
Identifiers: LCCN 2017027268 | ISBN 9781138646018 (hardback : alk. paper)
Subjects: LCSH: Emotions (Philosophy) | Risk assessment—Moral and ethical aspects. | Technological innovations—Moral and ethical aspects.
Classification: LCC B815 .R64 2017 | DDC 171/.2—dc23
LC record available at https://lccn.loc.gov/2017027268

ISBN: 978-1-138-64601-8 (hbk)
ISBN: 978-1-315-62780-9 (ebk)

Typeset in Sabon
by Apex CoVantage, LLC

To Jeff, Mae and Parker

Contents

Acknowledgments

I have worked on the topic of this book for more than ten years. The book incorporates thoroughly reworked and updated passages from the following articles that I have published on this topic: Roeser, S. (2013), 'Aesthetics as a Risk Factor in Designing Architecture', in Claudia Basta and Stefano Moroni (eds.), *Ethics, Design and Planning of the Built Environment*, Dordrecht: Springer, pp. 93–105 (included in Chapters 2 and 8); Roeser, S. (2007), 'Ethical Intuitions About Risks', *Safety Science Monitor* 11, pp. 1–30 (included in Chapter 3); Roeser, S. (2009), 'The Relation Between Cognition and Affect in Moral Judgments About Risk', in Asveld and Roeser (eds.), *The Ethics of Technological Risk*, London: Routledge/Earthscan, pp. 182–201 (included in Chapter 4); Roeser, S. (2006), 'The Role of Emotions in Judging the Moral Acceptability of Risks', *Safety Science* Vol. 44, pp. 689–700 (included in Chapter 5); Roeser, S. (2010), 'Emotional Reflection About Risks', Roeser, S. (ed.) (2010), *Emotions and Risky Technologies*, Springer, pp. 231–244; Roeser, S. (2014) 'The Unbearable Uncertainty Paradox', *Metaphilosophy* Vol. 45, Nos. 4–5, pp. 640–653 (included in Chapter 6); Roeser, S. (2012), 'Emotional Engineers: Toward Morally Responsible Engineering', *Science and Engineering Ethics* Vol. 18, No. 1 (included in Chapter 7), 103–115, Roeser, S. (2011) 'Nuclear Energy, Risk and Emotions', *Philosophy and Technology* Vol. 24, pp. 197–201 (included in Chapter 8), Roeser, S. (2012), 'Risk Communication, Public Engagement, and Climate Change: A Role for Emotions', *Risk Analysis* Vol. 32, 1033–1040 (included in Chapter 8) and reworked passages of my own contributions to the following co-authored papers: Sabine Roeser and Udo Pesch (2016) 'An Emotional Deliberation Approach to Risk', *Science, Technology & Human Values* Vol. 41, pp. 274–297 (included in Chapter 7), Jessica Nihlén Fahlquist and Sabine Roeser (2015), 'Nuclear Energy, Responsible Risk Communication and Moral Emotions: A Three Level Framework', *Journal of Risk Research* Vol. 18, No. 3, pp. 333–346 (included in Chapter 7) Jessica Nihlén Fahlquist and Sabine Roeser (2011), 'Ethical Problems with Information on Infant Feeding in Developed Countries', *Public Health Ethics* Vol. 4, pp. 192–202 (included in Chapter 8). I would like to thank the respective publishers and journals for their kind permissions to reuse this material.

My work for these articles was funded by a VENI grant on 'Emotions and Technological Risks' (grant nr. 275–20–007; 2005–2009) and by a VIDI grant on 'Moral Emotions and Risk Politics' (grant nr. 276–20–012; 2010–2015), both from the Netherlands Organization for Scientific Research (NWO), and by a NIAS-fellowship, at the Netherlands Institute for Advanced Studies (part of the Royal Dutch Academy of the Sciences, KNAW), with a project on 'Moral Emotions in the Design Process of Risky Technologies' (2009–2010).

I held these research grants while being employed at the Philosophy Department of TU Delft, at the Faculty of Technology, Policy and Management. I started to work on this book manuscript while being under the VIDI-scholarship, and I finalized it as part of my position at TU Delft. I am grateful to NWO, the NIAS and TU Delft for providing me with the opportunity and environment to develop my work over the last years. I am indebted to my colleagues at TU Delft for the great intellectual environment in which we continue to develop novel philosophical approaches for the challenges of the 21st century, and for the generous support of the university for our work.

I have presented the ideas of the articles and the book manuscript at numerous conferences and seminars. I am thankful to the audiences for their very insightful comments that have helped me to develop my ideas. I have also always learned a lot from discussions with students; our own PhD-students in philosophy of technology as well as the engineering students at all levels of study at TU Delft whom we provide with ethics teaching, and through my guest lectures at other universities, summer schools, etc. Furthermore, my research has benefited significantly from my work for various governmental advisory boards on which I served over the last years: the COGEM, a committee that advises the Dutch government about genetic modification, the advisory board of the COVRA/OPERA (Dutch research institute for nuclear waste disposal), and an advisory group on new risks for the Dutch ministry of infrastructure and environment (IenM). I have also presented my work to larger audiences at public events, such as TEDx Delft, and via interviews for public media. These interactions have always probed me to formulate my arguments as clearly as possible, and they have often provided me with inspiring ideas and critical questions. The work presented in this book has profited greatly from all these exchanges with scholars from a broad range of disciplines, policy makers and wider audiences.

I wish to thank Joanna Bouma for her thoughtful comments while editing the manuscript of this book; she has helped me yet again to express myself much more clearly and to make my thoughts more readable. I would like to thank Nathalie van den Heuvel for her kind assistance with compiling the bibliography. I wish to thank anonymous reviewers who provided me with helpful feedback on my draft book manuscript. I am grateful to the editors from Routledge, especially Margo Irvin, Andrew Weckenmann and Allie Simmons, for their support.

Last, I wish to thank my family for their encouragement and support for my research. My father has endowed my brother and me with an interest in engineering and technology ever since we were small children. My father and my mother have provided us with a love for books, music and culture. Both my parents have raised me with awareness for societal and ethical issues. This book reflects these roots, while I have taken my own turn on technology by emphasizing the role of emotions for ethical reflection and decision making on risks. My husband, Jeff Powell, has been a wonderful companion for the last two decades, both intellectually and emotionally. May our dear children, Parker and Mae, grow up in and contribute to a world that deals sensibly and sensitively with the challenges facing our world, and with risky technologies.

1 Introduction

Risk and Emotions

1.1 Introduction

Risky technologies often give rise to intense public debates. Many technologies are developed to improve human well-being. We largely owe our contemporary standard of living, with its high degree of sanitation and possibilities for travel, transportation and communication, to technology. Unfortunately, however, inherent to most technologies is also the chance of negative side effects or risks, such as pollution, accidents and superficial human relationships. Both the positive and negative aspects of technologies need to be assessed in order to develop improved technologies that achieve a more optimal balance between and distribution of risks and benefits. This assessment partially involves quantitative approaches, such as measuring the speed of an airplane, its harmful emissions per kilometer, the probability of a crash and so on. However, it also involves moral considerations. What value do we put on the efficiency of a technology as opposed to its possible disadvantages? What kinds of disadvantages should be measured, and how should they be balanced against each other? Which is worse, a technology with an average risk of one dead person per year or a technology with an average risk of five severely handicapped people per year? What is the value of a human life? How are risks and benefits distributed across society? Even though these questions require descriptive, empirical information, that information does not as yet constitute answers to the moral questions concerning risky technologies. These questions require moral reflection.[1]

While many contemporary risk scholars, from the social sciences, philosophy as well as from engineering, agree that moral reflection is essential in decision making on risky technologies, they think that such moral reflection should be rational and exclude emotions. This book challenges this dominant view and presents a novel approach to risk emotions, i.e., emotions that are evoked by or related to risk or risk perception. This book will argue that risk emotions are important ingredients to moral reflection on risky technologies.

Risk debates repeatedly end in stalemates between proponents and opponents, experts and laypeople. Typically, as emotions are supposedly irrational states that are immune to objective knowledge and reflection, emotions are blamed for creating or reinforcing these stalemates (cf. e.g., Cohen, 1998). This book provides an alternative thesis, namely that emotions should actually play a *vital* role in risk debates as they point to important moral values that need to be taken into account in decision making about risks. Emotions such as fear, enthusiasm, indignation and feelings of responsibility point to important moral values such as risks and benefits, as well as autonomy, justice and fairness. The latter three values in particular tend to be overlooked in conventional quantitative approaches to risk. Emotions are essential to moral reflection on risky technologies: they point out our values, and emotions such as sympathy, empathy and compassion help us critically reflect upon our own values and the values of others. By including relevant moral values and critical reflection, embracing emotions in decision making about risky technologies will lead to better decisions. At the same time, this approach allows greater understanding between stakeholders as it takes their views seriously. This new theory of risk emotions comprises epistemological, metaethical and normative ethical aspects as well as an approach to public deliberation with and about risk emotions. It is grounded in recent insights from risk scholarship and from the psychology and philosophy of emotions.

1.2 Risk Debates and Emotions

Many people are afraid of the potentially undesirable consequences of technologies. Empirical research has shown that people rely on emotions when making judgments concerning risks (Finucane et al., 2000; Slovic, 1999). Examples of risks associated with technological or scientific developments that spark heated and emotional debates include cloning, genetically modified (GM) foods, vaccinations, carbon capture and storage and nuclear energy. These debates repeatedly end in stalemates between proponents and opponents, experts and laypeople. Typically, emotions are blamed for these stalemates. While large numbers of the public are often afraid of the potential negative consequences of technologies such as these, experts typically assert that the risks are negligible. They often accuse the public of being emotional, irrational and immune to objective information. Policy makers usually respond to the gap between experts and the public in one of two ways: either by neglecting the emotional concerns of the public in favor of the experts, or by accepting the emotions of the public as an inevitable fact and as a reason to prohibit a controversial technological development. These responses are grounded in the assumption that the emotions of the public are irrational and block genuine debates.

The role of emotions in judging technological risk is the subject of empirical research, but it is little studied by philosophers. This is a pity because this issue involves pressing normative questions that cannot be answered by empirical research. This book aims to fill this gap by providing a normative philosophical discussion of the role of emotions in judging the moral acceptability of technological risks. The fundamental question is: can emotional responses to risk be warranted? Questions at a more practical level include: how we should deal with public emotional responses to risk; and should engineers, scientists and policy makers take the emotions of the public seriously when developing risk regulation?

In answer to these questions, the two dominant approaches in ethics would respond in different ways. Rationalists take emotions to be subjective and irrational; moral judgments should be made by reason. In terms of moral judgments about risk, rationalists would argue that the emotions of the public should be ignored because they are irrational. On the other hand, sentimentalists believe that ethics is grounded in subjective feelings and people's preferences. In the context of moral judgments about risks, they would argue that even though emotions are subjective, they should be a part of the decision-making process because they show us our preferences. However, since these preferences are seen to be merely subjective, they do not improve the intrinsic quality of our decision making. They only create democratic legitimacy and public support.

Furthermore, over the last decades, a lot of psychological research has been done on risk perception and decision making under uncertainty. The most influential school of thought is the one founded by Amos Tversky and Nobel laureate Daniel Kahnemann (Tversky and Kahneman, 1974; Gilovich et al., 2002; also see Kahneman, 2011 for a popularized account). This research provides empirical evidence that people do not process statistical information well, nor are they good in logical reasoning. Scholars working in this area argue that we have two distinct systems with which to process information, commonly called Dual Process Theory (DPT). According to the theory, people tend to rely on intuitive processes of thinking ('system 1') that provide them with fast heuristics that help them navigate smoothly through a complex world, but which are very unreliable. Analytical, rational thinking ('system 2') is more reliable, but it requires more time and resources. 'System 1' is supposed to be emotional, intuitive, unconscious and irrational, whereas 'system 2' is supposed to be analytical, deliberative, conscious and rational. This work is extremely influential in psychology and empirical decision theory, and recently more and more philosophers have become interested in this approach.

However, the assumption that emotions are irrational is far from obvious. To the contrary, many contemporary emotion scholars challenge the conventional dichotomy between reason and emotion that underlies rationalist and sentimentalist approaches as well as DPT. They argue that emotions are a form or source of practical rationality. This book aims to provide an alternative hypothesis to the dominant view about emotions in the literature on risk. It argues that emotions actually should play a more important role in risk debates as they point out important moral values that need to be considered in decision making about risks. This approach consists of a new epistemological theory of risk emotions, and of a new political philosophical theory on how to integrate risk emotions in decision making. The new approach is grounded in recent insights from the psychology and philosophy of emotions. In contrast to both rationalist and sentimentalist approaches in ethics, this book defends a cognitive theory of risk emotions according to which risk emotions are necessary to have *practical rationality* and are therefore of vital importance to decision making about risks. It argues that moral emotions are an indispensable normative guide in judging the moral acceptability of technological risks.

This book argues that this alternative view of emotions can lead to a different understanding of emotional responses to risk. Risk emotions can draw attention to morally salient aspects of risks that would otherwise escape our view. This alternative approach can shed new light on various controversial debates about risky technologies by showing that risk emotions can be reasonable. This does not mean that emotions are infallible; indeed, like all our mental resources, they are prone to errors and biases. However, emotions can themselves serve as a source of critical reflection by allowing us to take on other viewpoints via empathy and sympathy and through using our imagination. Such emotions can broaden our horizon and can allow us to adopt a critical stance toward our own initial responses. The new approach developed in this book proposes to take emotions as a starting point in debates about risky technologies, as they direct us to important moral values and concerns of all the people involved. This can lead to better decision making and at the same time, to greater understanding between stakeholders. By taking the emotions of the public seriously, the gap between experts and laypeople can eventually be overcome, leading to more fruitful discussions and decision making.

1.3 Overview of the Book

In this section, I provide for a short overview of the argument in the chapters to follow.

Part I, Risk Debates, Stalemates, Values and Emotions, provides for a critical discussion of current approaches to emotions, intuitions and moral values in decision making about risks.

Chapter 2. Emotions and Values in Current Approaches to Decision Making About Risk

Current approaches to decision making about risk can be classified under three general approaches: technocratic, populist and participatory approaches. All these approaches run into problems. I argue that this is due to an insufficient understanding of intuitive and emotional responses to risk, because these are seen as irrational states. *Technocratic approaches* to risk are based on quantitative methodologies such as cost benefit analysis. The technocratic approach neglects emotions, intuitions and moral values in favor of quantitative considerations. However, this means that important ethical aspects of risk are not acknowledged and the public is not involved, which contravenes the principles of democratic societies. *Populist approaches* take the intuitions and emotions of the public as the end point of discussions. While the public's negative emotions toward a risky technology may be considered irrational and an impediment to further discussion, for democratic reasons, the will of the public is taken as the decisive factor. While *participatory approaches* to technology assessment do grant the public a constructive role in decision making about risks, they either do not give explicit attention to emotions or emotions are seen as obstacles. This means that important concerns and values are not sufficiently acknowledged.

Chapter 3. Risk Perception, Intuitions and Values

This chapter reviews insights in public risk perceptions from empirical decision theory and how these can be interpreted philosophically. It argues that laypeople's risk perceptions frequently contain justified ethical intuitions. Empirical decision theorists study intuitive judgments under uncertainty, which depart from rational decision theory. Whereas some scholars, for example Nobel prize winner Daniel Kahneman, think that this shows that such judgments are unreliable, Paul Slovic and his colleagues show that laypeople have a different understanding of risk than experts. Their understanding includes qualitative aspects that do not play a role in quantitative approaches to risk. This chapter reviews the empirical findings by Slovic from a philosophical, normative perspective. It argues that laypeople's risk perceptions can be justified on philosophical grounds as being legitimate and rational. They often include important normative and evaluative considerations, such as justice, fairness, equity, and autonomy, considerations that do not play a role in standard, technocratic approaches to risk. Indeed, insights from risk ethics largely overlap with the insights of laypeople and social scientists. The chapter then provides a theoretical framework to understand laypeople's intuitive risk perceptions in a different way than psychologists and social scientists do. This framework is ethical intuitionism. Ethical intuitionism is a philosophical approach that takes intuitions as *prima facie* reasonable and justified direct perceptions of objective moral values. This chapter argues that risk perceptions, intuitions, and moral values are important for well-grounded judgments about acceptable risk. However, the chapter does not yet focus on emotions, as these add a layer of complication. The rest of the book focuses on risk emotions.

Part II, Reasonable Risk Emotions, develops an epistemological argument for why emotions should be taken seriously in decision making about risks. It argues that moral emotions can provide us with important insights into moral aspects of risk.

Chapter 4. Risk Emotions: The 'Affect Heuristic', its Biases and Beyond

Recent risk perception research focuses on the role of emotions. This research points out various ways in which risk emotions can be misleading, making most scholars hesitant about giving risk emotions an important role in debates. This chapter will analyze these arguments and identify some problems with them. The chapter starts with a discussion of the 'Affect Heuristic' approach by Paul Slovic and his colleagues. This approach holds that emotions shape people's risk perceptions. Slovic emphasizes that emotions can show us what we value, but that they are also prone to bias and need to be corrected by formal approaches. Slovic's views are largely indebted to the highly influential Dual Process Theory (DPT) developed by Daniel Kahneman and others. This approach sees emotions as irrational, unconscious intuitions and gut reactions that serve as heuristics in decision making under uncertainty but that are prone to bias. Although Slovic and Kahneman acknowledge the possibility that reason and emotion can interact, they view analytical approaches to risk as superior. I argue that although DPT provides us with important insights, it also gives rise to conceptual and practical problems, especially when it comes to emotions. I argue that emotions should be understood in a more nuanced way than suggested by DPT. Not all supposed biases in risk perception are real biases, and not all biases arise from emotions. A pressing practical problem is what I coin the 'Puzzle of Lay Rationality'. While Slovic's earlier work lends support to including laypeople in decision making about risks, his work on the affect heuristic threatens to undermine this by emphasizing formal methods as the final arbiter in decision making about risks. These problems can be solved by a different theory of emotions, which will be developed in the following chapter.

Chapter 5. The Philosophy of Moral Risk Emotions: Toward a New Paradigm of Risk Emotions

This chapter proposes an alternative approach to risk emotions, based on recent emotion research. Recent philosophical and psychological emotion research suggests that emotions are both affective and cognitive (so-called cognitive theories of emotions). They are a source of practical rationality and are important for moral knowledge. However, even though these ideas are common in emotion research, they are rarely considered in the literature on risk emotions. In the remainder of this book, I will show that these ideas can shed fruitful light on risk emotions. The idea that emotions can be a source of moral knowledge can be combined with the view of risk intuitions developed in Chapter 3. I argue that moral intuitions are paradigmatically cognitive moral emotions, and moral risk emotions are perceptions of ethical aspects of risk. This alternative view of risk emotions can shed new light on the emotion that has been the main focus of Slovic's studies, namely fear or dread. While fear can be irrational, it can also be a justified perception of danger. Furthermore, risk emotions also comprise paradigmatic moral emotions, such as sympathy, indignation and feelings of responsibility, which point out important moral values such as justice, fairness and autonomy. My alternative approach to risk emotions can solve the 'Puzzle of Lay Rationality': it is exactly because emotions play a role in laypeople's risk perceptions that laypeople have a broader approach to risk that includes important moral values.

Part III, entitled *Emotional Deliberation about Risk*, develops an approach to emotional deliberation about risk. This is a procedural approach to decision making about risks that takes emotional responses to technological risks, and the ethical concerns that lie behind them, seriously.

Chapter 6. Reflection on and with Risk Emotions

My alternative approach to risk emotions does not state that emotions are infallible. Risk emotions, and fear and disgust in particular, should be critically scrutinized. But emotions can themselves play a role in that process. Fear can be notoriously misleading, but fear can also point toward the morally problematic existential uncertainty that risks introduce, such as potentially large-scale catastrophic risks. Disgust can be an emotion that is biased toward a conservative status quo, but disgust can also point toward the morally ambiguous status of artefacts of synthetic biology, for example. I will discuss how we can correct misleading risk emotions. Emotions can themselves be a source of critical reflection and deliberation. I will specifically focus on sympathy and compassion to assist critical reflection on and the proper formation of reliable risk emotions.

Chapter 7. Participation with Emotion

My new approach to risk emotions builds on existing participatory approaches but provides for a new perspective by emphasizing that emotions should play an important role in decision making about risk as they point to important moral values. Rather than ignoring emotions as in the technocratic approach, or taking emotions as end points of discussion as in the populist approach, the new approach states that emotions should be the starting point of risk debates. It connects well with participatory approaches to risk while providing for an additional perspective. While such approaches are either implicitly or explicitly rationalistic, the approach developed in this book gives emotions an important role. By asking people what they are emotional about, substantive moral considerations underlying the emotions can be made explicit. This enables critical reflection on whether the emotions and considerations are justified. Not only laypeople but experts and policy makers too can be emotional about risks. However, contrary to technocratic approaches that try to avoid emotions in decision making about risk, the approach developed here welcomes the emotions of all stakeholders. Emotions can help experts and policy makers be aware of moral and societal responsibilities, and emotions can provide for mutual understanding of stakeholders. The new approach to risk emotions can lead to better decisions about risks, by increasing prospects to include all important moral values. Furthermore, it can also contribute to overcoming common stalemates by providing a framework that puts participants on an equal footing. This can provide for the willingness to give and take, to respect each other and to genuinely listen to each other.

Chapter 8. Emotional Deliberation on Technological Risks in Practice

This chapter discusses various emotionally charged debates about technological risks and how the emotional deliberation approach can shed fruitful light on these. The technological risks that are discussed are nuclear energy, as an example of an especially controversial technology, climate change, as an example of a systemic risk caused by our use of technologies, and which gives rise to major ethical issues on which emotions can shed important light, health technologies as examples of emotionally charged and sensitive topics with intricate ethical aspects, and architecture and urban planning as a technology domain that introduces another set of issues of which I argued that they should best be conceptualized as a further qualitative dimension of risk, namely related to aesthetics, and to which emotions can provide important insights. I then discuss aesthetics in a different context, namely concerning artworks that engage with risky technologies. I argue that these artworks can provide us with additional means to engage in emotional-moral reflection about risks.

I end with a short epilogue that wraps up the main insights of the book.

1.4 Conclusion

The book will develop philosophical arguments, but these will be grounded in thorough studies of empirical research into risk and emotion. This book provides a novel theoretical approach while at the same time offering concrete recommendations for decision making about risky technologies. It breaks with the current paradigm in risk scholarship that sees emotions as irrational states. Instead, this book will argue that specifically moral emotions are an important ingredient in decision making about risks, as they can highlight ethical aspects that get overlooked in conventional approaches to risk. A new understanding of risk emotions as sources of practical rationality can provide for a more fruitful approach to public decision making about risks than current alternatives.

A central feature of the arguments developed in this book is to try to stay as close to common sense as possible. It is inspired by a humanistic ideal that strives to do justice to our human capacities to make sense of the world. This goes against many contemporary approaches to human judgment that are essentially revisionist: they aim to show that what we believe are reliable sources of knowledge, are actually not reliable but are biased. While these approaches can heighten our alertness to our own fallibility, and in that sense contribute to critical thinking, they can also lead to cynicism about human judgment and eventually to a distrust in democracy.

My underlying intellectual adagio is as follows: let us try to see whether we can make sense of the empirical data based on philosophical approaches that are as charitable as possible to our human capacities, and let us only be wary of these capacities if all genuine and fair attempts to rehabilitate them have failed. As this book will show, despite the purported evidence of our cognitive and affective failures, these are not as extensive as some scholars have claimed, and they can give us hope in sustaining our efforts for democratic decision making.

My approach even aims to show that we should broaden democratic decision making to include aspects of our humanity that are often met with suspicion and neglect: our values, intuitions and emotions should play a vital role in democratic decision making. More specifically, this book argues that this should even be the case in a context where it might be highly tempting to exclude them, namely the context of decision making on technological risks. This is often seen as a domain in which expert judgment should take precedence over the broad public. Other scholars have already argued that this is based on a misunderstanding, as values and perceptions are unavoidable in decision making about risks. However, this book provides for an approach that goes further than these previous approaches, by providing a theoretical framework that shows that it is actually desirable to include values, intuitions and even emotions in decision making about risks, as only then can we fully appreciate the moral dimension of risks and come to constructive decision making that is genuinely democratic.

Note

1. As many anglophone philosophers, I use the notions 'moral' and 'ethical' interchangeably.

Part I

Risk Debates, Stalemates, Values and Emotions

Part I of the book will review current approaches to decision making about risks, and the role that emotions, intuitions and moral values play in these approaches.

2 Emotions and Values in Current Approaches to Decision Making About Risk

2.1 Introduction

A recurring pattern often emerges when a new technology is introduced: society is alarmed and worried about risks, while experts assure them that the risks are negligible. Policy makers often respond to these emotions in one of two ways: either they ignore the emotions of the public in favor of purely scientific information about risk, or they accept them and prohibit or restrict the technology in question. We can call these responses the 'technocratic pitfall' and the 'populist pitfall' respectively. I use the word 'pitfall' because in both cases, there is no genuine debate about the technology as the public is supposedly emotional and ill-informed and hence incapable of engaging in a rational debate based on objective, scientific information. This pattern has occurred in regard to nuclear energy, cloning, genetic modification, carbon capture and storage and vaccination, to mention just a few of many hotly debated, controversial technological developments. Stalemates such as these seem unavoidable if we assume that emotions are irrational and impenetrable by rational information. There is a third option, and that is to let the public actively participate in the assessment of the technology. This is an approach that has been successfully applied in several debates. However, here too, emotions are usually not explicitly addressed, as they are not seen to make a constructive contribution. This may mean that some of the concerns of the public are still not addressed.

These three approaches—technocratic, populist and participatory—of policy makers are also reflected in different approaches in risk scholarship. Quantitative approaches assess risk in a statistical way, with a focus on scientific and mathematical methods. This can be seen as a technocratic approach. Some scholars argue that because we live in a democracy, and because the public has different perceptions of risk, we should accept these different viewpoints and follow them. This can be seen as a populist approach, because there is no further debate about the perceptions of the public. The notions 'technocratic' and 'populist' are not typically used by scholars to refer to their own approaches, but I use them as labels to group certain types of approaches and to indicate their impact on policy. Next to these two 'ideal types' of approaches, there are participatory or social constructivist approaches in the social sciences. These emphasize both the differences of viewpoints amongst the public as well as the potential biases of experts. They also argue that we should engage the public via participatory approaches in decision making about risk.

In this chapter, I will briefly review these three ideal type approaches and argue that they are problematic given their understanding of values, intuitions and emotions as subjective, irrational or a-rational states. These approaches do not pay sufficient attention to responses to risk that, as I will argue, should actually play a role in decision making about risk. Furthermore, they are problematic in democratic societies as they do not do justice to stakeholders' important concerns. Participatory approaches do aim to include stakeholders and their values, but I argue that by not explicitly acknowledging the importance of emotions, these approaches do not yet do full justice to all the stakeholders' important concerns and values. In this chapter I will propose expanding participatory approaches with explicit attention for emotions. In the remaining chapters of this book, I will then provide for the theoretical underpinnings for including emotions in decision making about risk, for which broadening participatory approaches can be of major importance.

2.2 Technocracy

Conventional approaches to risk analysis and risk management are based on technocratic, formal methodologies. These approaches define risk as the probability of an unwanted effect. A typical approach to quantitative risk assessment is to determine the probability of annual fatalities as a consequence of a technology, for example through accidents or pollution. Policy makers then apply a cost benefit analysis (CBA, in the context of risk this is also often referred to as risk cost benefit analysis) to determine whether a technology should be implemented. CBAs measure the benefits of a technology in economic terms, for example, and balance these against possible negative side-effects or risks, calculated at an aggregate level by measuring expected overall utility. Proponents of this approach praise it as a rational, objective and value-neutral method (cf. e.g. Cross, 1998).

However, social scientists, empirical decision researchers and philosophers have criticized such formal approaches for oversimplifying the concept of risk and not doing justice to important qualitative and ethical considerations (Krimsky and Golding, 1992; Slovic, 2000; Hansson, 2004; Roeser et al., 2012). Risk cost benefit analysis leaves out important ethical considerations, such as responsibility, autonomy, justice, fairness and equity (Asveld and Roeser, 2009).

When assessing risk, we should look at two questions: one, what is the probability of unwanted consequences; and two, is the risk morally acceptable? The first question requires scientists and engineers to supply statistical data about risks and communicate them in a responsible way to the public and to policy makers. The second question, though, is a moral question. A function of probabilities and consequences is not sufficient to judge whether a risk is morally acceptable or not. Risk is not only a quantitative, factual notion, but it also involves values.

A standard CBA can be seen as a simple version of consequentialism. Consequentialism is a family of ethical theories according to which an action or a rule for actions[1] is right if it maximizes consequences on an overall, aggregate level. Non-consequentialist ethical theories such as deontology and virtue ethics object that this does not take into account important ethical considerations such as autonomy, justice, fairness, responsibility and agency. For example, we might achieve the best possible consequences with actions that exploit people against their will, or by violating promises. Non-consequentialist ethical theories such as deontology and virtue ethics state that these are important ethical considerations that need to be taken into account in their own right.

Analogously, CBAs do not typically include ethical considerations such as justice, fairness and autonomy. However, these are important ethical considerations that should be taken into account in decision making about risk. In chapter 3, I will provide a detailed argument concerning a range of ethical considerations that are relevant for decision making about risk but which are not included in quantitative approaches to risk such as CBA. Here I will highlight a few examples of such ethical considerations.

Based on the moral notions of autonomy or freedom, human beings should only be exposed to risks to which they have freely consented. For that reason, the principle of informed consent is crucial in, for example, medical treatments and research. Patients and research subjects should be free to participate in a medical trial or to take medication in full awareness of the possible risks of the treatment. This principle is also applied in other kinds of studies that involve human subjects. However, when it comes to technological risks, this can be difficult, as these are often collective risks that can affect large parts of the population, including people who have not consented to a risk, and who do not use a technology because they find it too dangerous (Asveld, 2008). Furthermore, people might have no choice but to accept specific risks if there are no alternatives available. This can be morally problematic. Another important ethical consideration is whether risks and benefits are fairly distributed. A risk-cost-benefit analysis assesses risks and benefits on an aggregate level. It does not look into who receives the benefits and who undergoes the risks, and whether the benefits and risks may be allocated fairly.

Hence, in order to decide whether a risk is *morally acceptable*, next to the balance of risks and benefits, the following moral considerations are important: autonomous decision making of those affected (informed consent); available alternatives; and a fair distribution of risks and benefits (for a more extensive list of moral considerations, see Chapter 3). Such considerations are not included in a CBA. This mirrors objections from deontology and virtue ethics against consequentialist approaches in ethics.

However, CBA is even more constrained than some versions of consequentialism. For example, utilitarianism requires the maximization of the well-being of *all* people (Sidgwick, 1901 [1874]), and some versions even consider other sentient beings (Bentham (2007) [1780]). Hence, utilitarianism has an altruistic goal. In contrast, a CBA can be constructed from the point of view of maximization of profits for a company or country. But even if it is aimed at increasing well-being for everyone, CBA inherits the problem of utilitarianism by not considering *how* this is achieved. A pure form of utilitarianism as well as a cost-benefit analysis would allow for the exploitation of minorities if this would lead to increased overall well-being. The end would justify the means. However, other ethical theories have criticized this implication of utilitarianism. For example, in one of his formulations of his 'categorical imperative', the philosopher Immanuel Kant has argued that we should never only use other people merely as a means, but always also as ends in themselves (Kant 1964 [1785]). This is a direct prohibition of an intentional exploitation of another person for the greater good, which can be the implication of a consequentialist assessment and a CBA. Kant's 'respect for persons' imperative also raises questions concerning quantifying the value of a human life in monetary terms, as is frequently done in cost-benefit analyses.

Hence, far from being value neutral, CBA is inherently value-laden by being a covert form of consequentialism, with its evaluative commitments to maximizing aggregate well-being and excluding non-consequentialist considerations. However, as argued above, the implicit evaluative assumptions are highly contestable. Furthermore, ethical considerations also play a role in risk assessment and in a CBA when determining the kinds of effects to take into account. For example, should only annual fatalities be taken into account, or also the number of sick or injured people or effects on nature?

Various philosophers of risk argue that, on top of the ethical problems that relate to disregarding issues of distribution, cost benefit analysis faces serious methodological problems (Hansson, 2004; Shrader-Frechette, 1991; contributions to Asveld and Roeser, 2009). The methodological problems concern questions of how to measure and compare different sorts of well-being and how to value a human life. It is difficult or even impossible to express all moral considerations about technologies in terms of risks or costs and benefits and to compare them on one scale (Espinoza, 2009). The methodological assumptions in technocratic approaches to risk are often highly arbitrary while potentially making a huge difference to the comparative assessment of various risky activities. For example, depending on the value placed on a human life in a cost benefit analysis, the outcomes can be diametrically opposed. Cost benefit analysis gives us an illusion of objectivity, blurring the underlying substantial ethical and methodological considerations rather than making these explicit and subject to critical deliberation.

This gives rise to the question of how to balance the different ethical considerations in a risk assessment. Where cost benefit analysis apparently does provide us with a quantifiable methodology to risk, alternative approaches do not provide clear-cut answers on how to weigh different ethical considerations. For example, we may need to make potential trade-offs in equity and fairness versus overall, aggregate well-being. Some propose designing models that give weight to particular factors and values (cf. e.g., Bohnenblust and Slovic, 1998). Although it is possible to design a quantitative model that incorporates values and trade-offs between values, the philosophical question remains whether these trade-offs are morally justified and whether the same trade-offs can be made in each and every case. This relates to a much more fundamental problem. It is unclear whether potentially conflicting ethical considerations can be balanced in abstraction of concrete circumstances. This question is a topic of major debate in metaethics. Virtue ethicists, particularists and other defenders of 'context-sensitive' approaches argue that how to balance different ethical considerations must be judged on a case-by-case basis (Dancy, 2004; McDowell, 1998). I will discuss this in more detail in Chapter 3.

Furthermore, as Cross (1998) argues, even potentially legitimate values can also be biased. Cross sees this as a reason to be careful with including values, and favors a quantitative approach instead. However, one should not extrapolate examples of misguided values to all values. As I will argue in Chapter 3, values are inevitable. Even if people would be mistaken about values most of the time, this would be an empirical fact (albeit in need of normative assessment), but it would not provide for a normative argument. If people would be mistaken about values most of the time, this would not have bearing on the normative weight of values—that is, it would not undermine the normative validity of values as such. All potentially reasonable considerations can be biased as well, but rather than dismissing them across the board, we should critically assess and reflect on them, a process that requires context-sensitive insights and deliberation that cannot be replaced by a formal, predetermined methodology.

Of course, decision making about large-scale risks requires quantitative and statistical information. However, one should always be aware of their limitations. Scientific methods are necessary to assess risks, and cost-benefit analysis can play an important role in comparing risks, but these approaches cannot *replace* a genuine ethical assessment. Rather, they are prerequisites that should *inform* an ethical assessment. Modeling moral trade-offs *requires* explicit moral reflection; it cannot *replace* it. Modeling moral trade-offs should then at most be a tool for ethical reflection in an iterative process. Risk decisions are inherently morally complex. It is important to recognize and face this moral complexity, even if it means that one cannot fall back on a given, clear-cut methodology. Assessing the moral acceptability of risks will inevitably involve contextual, situation-specific deliberation. Such a process of deliberation should involve different stakeholders, as they can provide for a broad range of ethical insights. These ideas will be elaborated on in much more detail in the remainder of this book.

For now, we can observe that the technocratic approach neglects emotions, intuitions and moral values in favor of quantitative considerations. However, that means that important ethical aspects of risk are not acknowledged and that the public is not involved, which is problematic in a democratic society.

2.3 Populism

In technocratic approaches, risk assessment is a matter of statistical, quantitative information, leaving no room for intuitions, emotions and moral concerns of stakeholders. In contrast, in populist approaches, the emotions and concerns of the public are taken for granted and are seen as inevitable. If there is no public support, a risky technology cannot be implemented. There are also scholars who defend the populist approach for instrumental reasons, namely to ensure societal support for technological developments by appealing to people's emotions (De Hollander and Hanemaaijer, 2003).

Technocratic and populist approaches rest on the assumption that reason and science are categorically distinct from emotion and perception, with the latter being inferior. I will come back to this in more detail in Chapter 4, where I will critically examine this supposed opposition and hierarchy. Endorsing this opposition and hierarchy leads both the technocratic as well as the populist approaches to the view that the risk perceptions of the public cannot be critically scrutinized as they are by definition mistaken. While technocratic approaches reject the public's risk perceptions for that reason, populist approaches argue that these risk perceptions should be respected on democratic grounds.

Renn (1998) describes this kind of approach as one in which risk communicators and risk managers draw on psychometric studies of public risk perception and then attempt to address the dominant values and concerns that emanate from these studies. These studies tend to be performed on a high level of aggregation. However, Renn discusses that it eventually became clear that there was a large variation and diffusion underlying these seemingly homogenous risk perceptions. This gave rise to methodological and practical problems and implications. He writes:

> Which estimate should then be used for risk management purposes? The average of all groups, the mean value of all respondents with college degrees, the average of all women (since they tend to respond more cautiously than men to most technological risks)? The gap between experts and the public has been transformed into numerous gaps among experts and among publics (Fischhoff, 1996, p. 83). Being confused by this variety, many risk managers have abandoned the idea of public input altogether and opted to return to the safe haven of institutional or technical expertise.
>
> (Renn, 1998, p. 51)

What Renn discusses here is a problematic aspect of the populist approach, namely how to identify who the public is and which values the public endorses. As Renn and other social scientists have argued, the public is not homogenous and does not endorse just one view. Rather, there are numerous publics with numerous cultural and evaluative outlooks. These differences within the public get overlooked if one only focuses on risk perceptions on an aggregate level. Renn describes how this can result in risk managers taking recourse to technocratic approaches, because they seemingly provide clear-cut solutions. Hence, in the populist approach, the views of the public are taken as endpoints of debate, but when matters turn out to be more complex, the populist approach may get traded in for a technocratic approach, or vice versa.

Sometimes the technocratic and the populist pitfalls occur in the same debate, for example, in the case of the debate about carbon capture and storage (CCS) in the Netherlands. Initially, a CCS facility was to be built in Barendrecht. Experts told the concerned public that there were no risks, and the public was not involved in the decision making (cf. Feenstra et al., 2010 for a detailed discussion of this case). The concerns of the public were dismissed as being emotional and irrational. This can be seen as an instance of the technocratic pitfall. However, resistance was so strong that the initial plans were abandoned and rescheduled to the much less densely populated province of Groningen. However, the people there also rejected these plans from the start: why should they accept something that other people did not want in their backyard? Groningen was then already starting to suffer from earthquakes resulting from natural gas extraction, and the population there argued that they did not want to be the 'drain' ('afvoerputje') of the Netherlands.[2] Politicians quickly gave up the plan because there was apparently no 'social support'.[3] This can be seen as an instance of the populist pitfall: the will of the public is taken to be definitive, with no attempt at a genuine dialogue on the pros and cons. However, the divergence of public views could also be taken as a starting point for discussion and deliberation on risky technologies, in order to achieve more responsible innovations. This is an idea that we will see when we discuss participatory approaches in the next section.

What I discussed here concerns the way populist approaches deal with *values*. We can see parallels in the way populist approaches deal with *emotions*. The populist approach takes the emotions of the public as the end point of discussion. If the public is emotional and against a risky technology, this is considered as a position that does not allow further discussion. Scholars who write on risk and feeling also argue along similar lines. For example, Loewenstein et al. (2001, p. 281) argue as follows:

> Simply disregarding the public's fear and basing the policy on the experts, is difficult in a democracy and ignores the real costs that fears impose on people, as is well documented in the literatures on stress and

> anxiety. The best policy, then, would be one that involves mitigating real risks and irrational fears.

However, in this case too, there is no genuine deliberation or public debate about emotions, concerns and values, as such a debate is considered to be impossible. 'Real risks' are here juxtaposed with 'irrational fears'. Because of their potential cost implications, the emotions of the public are only considered instrumentally. That emotions might point to reasonable considerations that might be valuable contributions to decision making is not taken into account. This is also reflected in the model that Loewenstein and his colleagues present. The model lists the following antecedents of emotions: "anticipated outcomes (including anticipated emotions)," "subjective probabilities," and "Other factors e.g. vividness, immediacy, background mood" (Loewenstein et al., 2001, p. 270). These antecedents are not directly relevant for a better understanding of risk. Subjective probabilities may not be warranted, and vividness, immediacy and background mood can influence risk perceptions without being well-grounded. Anticipated emotions are mentioned in the case of anticipated outcomes, but these do not need to reflect any justified concerns.

In another publication, Loewenstein and Lerner (2003) distinguish between two different types of affect or emotion that are relevant for decision making. The first type is 'immediate affect' that is, direct responses that influence a decision. The second type is 'anticipated affect', that is, foreseen emotional impact of expected outcomes. They discuss potential biases and benefits of both types of emotions. A risk of immediate affect is that it can lead people astray and cloud their judgments, but benefits can be: 'Prioritizing information processing and introducing important, but intangible, considerations' (Loewenstein and Lerner, 2003, p. 634). Zinn (2008) makes a similar point when emphasizing that emotions can attract our attention and in that way prevent harm, and they can be an 'immediate resource to guide action' and serve as an 'alarm bell' (Zinn, 2018, p. 447). Anticipated affect, on the other hand, can help to make decisions that maximize well-being, but if expectations are biased, so will be decision making. However, note that biases concerning expectations do not need to be due to emotions; they can also be based on false factual information. In any case, in this publication, Loewenstein and Lerner point to possible positive contributions of emotions to decision making, next to discussing potential biases. I will come back to a thorough discussion of (alleged) emotional biases to decision making under risk in Chapters 4 and 6.

Another possible class of antecedents of emotions are values. These could, for example, play a role in the context of anticipated outcomes. Loewenstein and his colleagues and other scholars who write on risk and emotion usually do not mention the role of values. Yet, a common insight from emotion research in philosophy and psychology is that emotions are typically responses to values, or responses to situations that affect one's important values. This insight would justify taking emotions seriously in debates about risk as a way to shed light on people's important values. However, this has not been discussed in the work of most scholars who write about risk and emotion (although cf. Kahan, 2008). This means that an important alternative to technocratic and populist approaches to risk and emotion is not considered in the academic literature.

In the next section, I will discuss participatory approaches. By explicitly addressing different perspectives and values of stakeholders, these are promising alternatives to technocratic and populist approaches. However, participatory approaches do not usually explicitly address emotions, which means that important concerns may not be taken into account (Roeser and Pesch, 2016).

2.4 Participation

Many social scientists and philosophers working on risk have identified the shortcomings of the technocratic and populist approaches. They have pointed out the false dichotomies between facts and values in the context of risk (cf. e.g., Mayo and Hollander, 1994), as well as between experts and laypeople (Jasanoff, 1998). As I discussed in the section on technocracy, these scholars emphasize the inherent value-ladenness of quantitative approaches to risk. Many social scientists also discuss the values and potential biases of experts that color their approaches, which undermine their claimed objectivity and neutrality. While the assumption of technocratic and populist approaches is that experts have a superior approach to risk, many social scientists who study risk emphasize that different stakeholders have different but equally legitimate perspectives on risk.

Facts and values are often highly contested and are constructed differently by different stakeholders and cultural groups (Krimsky and Golding, 1992; Slovic, 2000; Kahan, 2012). Participatory approaches to technology assessment aim to involve all stakeholders in decision making about risk in order to do justice to these different perspectives. Such approaches grant the public a constructive role in decision making about risks, not just for reasons of majority vote or for pragmatic or instrumental reasons, as in populism, but in many cases because it is assumed that laypeople can make important substantive contributions to decision making about risks. For example, Cuppen (2012) argues that in order to genuinely learn from each other, constructive conflict, rather than settling too early for a consensus that leaves out important issues that deserve attention and deliberation, can be a promising approach to stakeholder dialogue.

The justification of participatory approaches is also based on a democratic ideal: people should have a genuine say in decision making about risky technologies. The moral principles of autonomy and equality decree that the moral views of the public should be taken into account in decision-making procedures concerning the implementation of risky technologies. Shrader-Frechette

(1991) provides for a philosophical exploration on how to improve risk assessment methodologies and procedures in order to make them more democratic and do justice to the fact that risk assessment is an inherently moral endeavor.

Participatory approaches include, for example, consensus conferences and town meetings (cf. e.g., Gregory and Keeney, 1994; Sclove, 2000; Jaeger et al., 2001; McDaniels et al., 1999; and van Asselt and Rijkens-Klomp, 2002, for discussions of different participatory methods). Correlje et al. (2015) argue that participatory approaches can contribute to 'responsible innovation' by incorporating stakeholders' values via 'value sensitive design.' I will come back to these notions in the Chapters 7 and 8 of this book, where I will argue that emotions can be an important source of insight into the values of stakeholders in public debates about responsible innovation of risky technologies. However, in standard participatory approaches, emotions are not mentioned or are even explicitly denied a constructive role (cf. Harvey, 2009; Roeser and Pesch, 2016 who criticize participatory approaches for this; also cf. Chapter 7 for a more detailed discussion of this).

Many social scientists who study risk take values to be subjective states (e.g., Jasanoff, 1998). They argue that the distinction between objective facts and subjective values is mistaken. Rather, both are intertwined and, according to social scientists, for that reason subjective. However, this ultimately means that there are no better or worse evaluative perspectives. Yet, it is philosophically far from obvious that values are purely subjective. There is an ongoing debate in metaethics as to the subjectivity or objectivity of ethics. There are good philosophical arguments to understand values in an objective way, without reducing them to scientific facts. Rather, values concern different aspects of objective reality than scientific facts. Arguing that evaluative risk judgments are subjective can undermine their credibility, as it can suggest that all risk perceptions are arbitrary (cf. Roeser, 2006; Möller, 2012; Hermansson, 2011 for detailed arguments against taking evaluative aspects of risk as subjective). However, we can distinguish between more and less legitimate evaluative risk judgments that can be more or less normatively justified. Although facts and values can be intertwined in the context of risk (cf. Möller, 2012), this does not mean that they are subjective, and we can still conceptually distinguish between them and assess them in different ways. We can assess scientific aspects of risk using scientific methods and moral aspects of risk using ethical reflection. I will discuss this in more detail in Chapters 3 and 4. Value conflicts cannot be solved by scientific data alone. As I will argue in Chapters 7 and 8, a deliberation between stakeholders can provide for better understanding of different perspectives via imagination and sympathy, as well as for a genuine exchange of arguments, which can lead to revisions of one's own perspectives, values and emotions.

Furthermore, these approaches either do not pay explicit attention to emotions, or they see emotions as obstacles to decision making, because they are taken to be irrational, subjective or arbitrary (cf. Roeser and Pesch,

2016). As I will argue in the main body of this book, this is a major omission, as emotions can play a meaningful role in pointing out important moral values in the context of risk. Rather than dismissing or evading emotions, they should be explicitly addressed in order to highlight values, provide insight in each other's perspectives and thereby contribute to more thorough moral reflection as well as to mutual understanding.

Hence, while I think that participatory approaches to risk assessment are the right way to go, the theoretical justifications and underpinnings of these approaches can be strengthened in two respects, namely concerning their views on values and on emotions, which also has implications for their practical applications in policy.

I will argue in the chapters to follow that there are good philosophical grounds to understand values in the context of risk in an objective, albeit non-reductive way, and to take emotions as rational or reasonable states that provide us with epistemic access to values in the context of risk. These interpretations of values and emotions lend much stronger support to including stakeholders in public decision making about risk than populism and conventional participatory approaches do, which in turn can provide for more solid foundations for participatory policies. This is what I call the 'emotional deliberation approach to risk'. Furthermore, by explicitly paying attention to people's emotions, this approach provides foundations as well as guidelines on why and how to constructively incorporate values of stakeholders in decision making about risky technologies.

Jaeger et al. (2001, pp. 159–165) describe the successful application of what they call the 'Cooperative Discourse model' in Switzerland at the beginning of the 1990s. Jaeger et al. (2001) discuss the limitations of the view of rationality underlying mainstream risk analysis. Risk analysis is based on normative decision theory, which does not fully correspond to the way people actually make decisions. Jaeger et al. argue that we should employ a broader notion of rationality in decision making about risk. My argument in this book is that emotions should be included in this type of broader, and more appropriate, notion of rationality.

In Table 2.1 below, I provide a crude overview of the different views on the status of facts, values, emotions and stakeholders for each of the ideal-typical approaches that I have discussed in this chapter.

Note that these labels simplify intricate debates and that there are also hybrid and more nuanced accounts. Nevertheless, this overview can help to distinguish the ideal-typical approaches that I discussed earlier in this chapter.

In Chapter 3 I will discuss the ethical intuitions underlying laypeople's risk perceptions, and I will argue that these can represent justified moral concerns that are an important addition to quantitative approaches to risk. The rest of the book will focus specifically on emotions in the context of risk. In Chapter 4 I will critically examine the theoretical underpinnings of theories that see emotions as a threat to decision making about risk. In Chapter 5,

Table 2.1 Facts, values, emotions and stakeholders in ideal-typical approaches to risk

	Facts	*Values*	*Emotions*	*Stakeholders*
Technocracy	Objective	Subjective	Irrational	Exclude
Populism	Objective	Subjective	Irrational	Include
Participatory Approaches	Subjective	Subjective	Irrational	Include
Emotional Deliberation Approach	Objective	Objective	Rational	Include

I will argue that the common view of emotions as purely subjective and irrational states has been challenged by emotion researchers. In Chapter 5, I will present an alternative approach to emotions in the context of risk, namely as sources of practical rationality and insight into the moral dimension of risk. In Chapter 6, I will discuss how we can critically deliberate on and with emotions, in order to critically assess them and to distinguish biased emotions from reasonable emotions. In Chapter 7, I will develop a so-called 'emotional deliberation approach to risk.' This approach can be seen as a direct supplement to participatory approaches to risk assessment as discussed in this section. In Chapter 8, I will discuss a few examples of hotly debated risky technologies, and how the approach developed in the previous chapters can provide for new insights in how to improve these debates by paying explicit attention to emotions and underlying values.

2.5 Conclusion

In this chapter I have briefly reviewed the dominant approaches to decision making about risk, which I have labeled technocratic, populist and participatory. I have argued that these approaches neglect emotions. They do this explicitly in technocratic approaches, where they are denied a contribution altogether; or they take emotions as irrational end points of debates, as in populist approaches, while in participatory approaches they are either not mentioned or are even explicitly denied relevance. The approach I will develop in this book rejects the dichotomy between reason and emotion and offers an alternative to these approaches. Emotions can be a source of moral reflection and deliberation. They allow us to avoid the technocratic and the populist pitfalls, and to enrich current participatory approaches. I argue in favor of an 'emotional deliberation approach' to risk, by integrating emotions into participatory approaches. Emotions should play an explicit role in debates and public deliberations about risky technologies, in the process of which people discuss their underlying concerns. Rather than ignoring them as in the technocratic pitfall or taking them to be end points of debates as in the populist pitfall, emotions should be taken seriously. They may reveal genuine ethical concerns that are important; they may also show biases and irrationality that can be addressed by information that is presented in an emotionally accessible way. They may also challenge one's own narrow perspectives by broadening one's imagination via sympathy, compassion and feelings of responsibility, for example.

I will argue in this book that taking emotions into account can enable a genuine dialogue and lead to morally better decisions. As a side effect, this approach can also contribute to a better understanding between experts and laypeople. There will probably be a greater willingness to give and take if both parties feel they are taken seriously. While this procedure might seem more costly, it is likely to be more effective, and hence more fruitful, in the long run. Currently, many debates about risky technologies result in an even wider gap between proponents and opponents and in rejections of technologies that could make a positive contribution to society if developed and introduced in a morally sound way. Genuinely including emotional concerns in debates about risky technologies can help overcome such predictable stalemates and lead to more responsible innovations. These are ideas that I will develop step by step in the chapters to follow.

Notes

1. These are act consequentialism and rule consequentialism respectively.
2. www.trouw.nl/home/-noorden-mag-geen-afvoerputje-worden-adde8ec3/
3. www.volkskrant.nl/politiek/kabinet-ziet-af-van-co2-opslag-in-noorden-a1834744/

3 Risk Perception, Intuitions and Values

3.1 Introduction

In the previous chapter, I reviewed the dominant approaches to risk assessment, which I labeled technocracy, populism and participation. I indicated that the last approach tries to engage the public in decision making about risky technologies, not just for pragmatic or instrumental reasons such as populism, but in many cases because it is also assumed that laypeople can make an important contribution to decision making about risks. In order to appreciate and justify this approach, it is helpful to review an important debate in empirical decision theory on differences in risk perceptions between experts and laypeople. The chapter will then provide a novel philosophical approach to interpret the findings from this empirical debate that provides even more substance to including laypeople in decision making about risk.

Psychological and sociological research shows that laypeople have a different understanding of risk than experts, involving intuitions, value judgments and context-sensitive features. Risk scholars view aspects of risk perception such as these as subjective (Finucane et al., 2000; Krimsky and Golding, 1992; Slovic, 1999). Several scholars also argue that risk is a social construction because people have different notions of risk (Slovic, 1992, 1999; many authors in Krimsky and Golding, 1992 express this view). The question is whether these interpretations really help to improve the debate on acceptable risks. The danger of taking value judgments as subjective and risk as a social construction that does not really exist is that concerns about risks can be ignored as supposedly being irrelevant or maybe even misguided (Shrader-Frechette, 1991). This is indeed argued by Weiner (1993). Many scholars think that intuitive judgments about risks are a bad guide (Gilovich et al., 2002).

However, whether intuitive risk judgments concern quantitative aspects of risk, or whether intuitive risk judgments concern the acceptability of risk can be ambiguous. When judging risks, we can ask two very different kinds of questions.

1. How great is a risk quantitatively; how likely is it to happen; what are the presumed consequences?

And:

2. How acceptable is a risk?

While the first quantitative risk judgment involves applying scientific methods, the second judgment by definition requires normative considerations: when is it justified to impose dangers on others? And how should we judge whether a risk is morally acceptable or not? Social scientists and ethicists working on risk agree that these normative questions cannot be reduced to the quantitative analysis of risk (Fischhoff et al., 1981; Shrader-Frechette, 1991; Roeser et al., 2012). This chapter reviews important developments in the psychological and philosophical study of acceptable risk. It will propose a new account that allows for the idea that intuitions about the evaluative aspects of risk are not subjective but can be reasonable concerns that should be taken seriously. My argument is based on the metaethical theory of ethical intuitionism, but I will expand this theory to the context of risk. I will argue that whereas intuitions are an unreliable source of insight for quantitative aspects of risk, intuitions might be inevitable and legitimate when it comes to judging normative, ethical aspects of risk.

3.2 Risk Perception Studies and 'Lay Rationality'

Research into public risk perceptions started in the late 1960s and early 1970s with the rise of empirical decision theory. Most famously, Tversky and Kahneman (1974) have shown how people make judgments on risks that diverge significantly from judgments based on normative decision theories such as rational choice theory. Rational choice theory, expected utility theory and other rational decision theories had until then been the academically dominant approaches, and they were based on formal rules and methods (cf. Peterson, 2009; and part III of Roeser et al., 2012 for introductions, overviews and critical discussions of rational decision theory). It transpired that not only laypeople, but experts, too, take decisions in ways that deviate from these strict rules and methods, and that they too have problems processing statistical information (Tversky and Kahneman, 1974; also see Gigerenzer, 2002). This gave rise to a whole industry of investigations into the biases to which people are prone in risk judgments, under the header 'heuristics and biases'. This research would eventually result in a Nobel Prize in Economic Sciences for Daniel Kahneman in 2002 (cf. Kahneman, 2011 for a popularized summary of his research).

Many empirical decision researchers think that intuitive judgments are an unreliable guide in decision making about risks. They argue that although intuitive judgments might have pragmatic value in allowing us to navigate through a complex world (in that sense they are heuristics), we should not expect them to correctly represent probabilities or relations between probabilities (in that sense they are distortions or biases; cf. the title *Heuristics and*

Biases: The Psychology of Intuitive Judgement by Gilovich et al., 2002). The following are examples of how intuitive judgments concerning probabilities can lead us astray:

1. 'the law of small numbers': overgeneralization on the basis of small samples;
2. judgments of causality and correlation: perception of relationships that one expects to see, even if they are not there;
3. availability: influence of imaginability or memorability on the perception of hazards;
4. anchoring and insufficient adjustment: making use of simple starting points and adjustment mechanisms;
5. information processing shortcuts: using simple decision strategies that avoid the weighing of multiple considerations;
6. probability neglect: focusing on the worst case even if it is highly improbable;
7. loss aversion: dislike of loss of the status quo;
8. 'benevolence' of nature versus 'suspect' man-made processes;
9. system neglect: inability to see that risks are part of systems, where intervention in a system can create risks;
10. 'framing': the way information is presented influences the choices people make.[1]

Sunstein (2005)

However, there is an important ambiguity when it comes to judgments about risks. These judgments may concern the purely quantitative risk (in terms of the probability of a given effect), but they might also concern the moral acceptability of a risk. The two are not identical (cf. for example, Fischhoff et al., 1981; and Shrader-Frechette, 1991). It is possible that in judging risks, people not only assess quantitative aspects of risk but at the same time, assess the desirability of the risks. The latter is a normative issue.[2] There is evidence that risk judgments by laypeople indeed incorporate both aspects (Slovic, 2000, 116). If this is true, then the fact that intuitive judgments diverge from probabilistic judgments does not necessarily mean that intuitive judgments are flawed. It might mean that in intuitive judgments, we can capture more than mere probabilities can tell us.

Since the 1970s, Paul Slovic and his colleagues have conducted numerous psychometric studies into the risk perceptions of laypeople. This research began with the assumption that insofar as risk perceptions deviate from rational decision theory, they are biases. Slovic's psychometric paradigm was for this reason heavily criticized by sociologists (cf. Jasanoff, 1998). However, Slovic eventually started to develop an alternative hypothesis, namely, that laypeople not so much have a wrong perception of risk, but rather they have a different perception of risk than experts. This change in Slovic's work is insufficiently acknowledged by, for example, Jasanoff (1998), who only seems to refer to Slovic's earlier work. This alternative hypothesis was supported by Slovic and his colleagues' findings that if asked to judge *annual fatalities* as a result of certain activities or technologies, laypeople's estimates came close to those of experts. However, when asked to judge the *risks* of a certain activity or technology, laypeople's estimates differed significantly from those of experts. Experts define risk as the probability of an unwanted effect and most commonly, as annual fatalities, so they see the two notions as by and large the same. However, apparently for laypeople, these are different notions. They seem to have different connotations of the notion of risk that go beyond annual fatalities:

> These results lead us to reject the idea that laypeople wanted to equate risk with annual fatality estimates but were inaccurate in doing so. Instead, we are led to believe that laypeople incorporate other considerations besides annual fatalities into their concept of risk.
>
> (Slovic, 2000, 116)

Slovic and his colleagues conducted studies with which they tried to disentangle the additional considerations that play a role in laypeople's risk perceptions. They eventually developed a list of 18 additional considerations (Slovic, 2000, p. 86) and their inter-correlations (Slovic, 2000, 140):

1. Severity not controllable
2. Dread
3. Globally catastrophic
4. Little preventive control
5. Certain to be fatal
6. Risks and benefits inequitable
7. Catastrophic
8. Threatens future generations
9. Not easily reduced
10. Risks increasing
11. Involuntary
12. Affects me personally
13. Not observable
14. Unknown to those exposed
15. Effects immediate
16. New (unfamiliar)
17. Unknown to science
18. Many people exposed

(Slovic, 2000, 140)

These characteristics were found to be important features in laypeople's risk judgments in previous studies by Slovic and other researchers (Slovic, 2000, 86). Similar characteristics have been mentioned by other scholars as well (cf. e.g., Renn, 1998). This empirical material gives rise to the normative, philosophical question whether the concerns of laypeople are justified or reasonable. Slovic thinks that this is indeed the case:

> Perhaps the most important message from the research done to date is that there is wisdom as well as error in public attitudes and perceptions. Laypeople sometimes lack certain information about hazards. However, their basic conceptualization of risk is much richer than that of experts and reflects legitimate concerns that are typically omitted from expert risk assessments.
>
> (Slovic, 2000, 191)

3.3 Risk, Values and Decision Making

Slovic thinks that laypeople have a richer perception of risk than experts, because it reflects legitimate concerns. This is a normative claim that cannot be established by empirical research alone. In this section, I will discuss Slovic's normative claim in more detail by providing a normative analysis of the characteristics that Slovic has found. But let us first again consider how experts understand risk (also cf. chapter 2).

In the standard approach to risk assessment and risk management, risk is understood as a function of probabilities and unwanted outcomes. Conventional approaches to risk assessment are based on quantitative methodologies such as cost benefit analysis (CBA). Examples of unwanted consequences are the number of deaths or injuries, or the degree of pollution. Policy makers use CBA to weigh the possible advantages of a technology against its possible disadvantages. They praise CBA as an objective and value neutral method.

However, risk and safety are inherently normative notions, or so-called 'thick concepts': they have both factual and ethical aspects at the same time (Möller, 2012). Moral philosophers emphasize that one cannot simply derive values from facts (Hume, 1975, [1739–1740]; Moore, 1988 [1903]). The quantitative, scientific aspects of risk are studied by empirical disciplines, but the evaluative aspects of risk require ethical reflection. Risky technologies can affect people's well-being. Determining how to balance the value of human life, long-term illnesses and environmental effects, and how to distribute risks and benefits cannot be done by purely quantitative methods. It also involves ethical reflection that goes beyond conventional approaches to risk such as CBA (Asveld and Roeser, 2009). CBA resembles utilitarian or consequentialist theories in ethics, which state that we should maximize aggregate benefits or minimize unwanted outcomes. Note that determining what counts as 'unwanted outcomes' already involves an ethical judgment and cannot be made on purely quantitative grounds (Fischhoff et al., 1981; Jasanoff, 1993; Slovic, 1999; Stirling, 2002). Furthermore, utilitarian approaches are subject to severe criticism in moral philosophy. Common objections against pure utilitarianism are that it ignores issues of fair distribution, justice, autonomy and motives by purely focusing on the consequences of actions. Many contemporary utilitarians try to include these considerations in their approaches,[3] but CBA is based on more simplistic utilitarian approaches and hence does not do justice to these ethical considerations.

Let us now examine whether the risk considerations that Slovic found to play a role in laypeople's risk perceptions can be seen as justified moral concerns, and whether they are or can be included in quantitative approaches to risk analysis such as CBA.

'Severity not controllable' (1): this is clearly morally relevant. Allowing an activity with severe potential negative effects that are beyond our control is like opening Pandora's box. This is a risk that we might only be prepared to take if much can be gained that cannot be gained in a better way.

'Dread' (2):[4] it is an open question whether something that is frightening should be taken seriously. It seems reasonable to say that it depends on whether one is justified in perceiving something as frightening.[5]

'Globally catastrophic' (3): the full text of the question refers to 'catastrophic death and destruction across the whole world'. It seems self-evident that it is a legitimate moral concern to take into account whether a risk might cause a global catastrophe.

'Little preventive control' (4): it is relevant for the moral acceptability of a technology or activity whether we may or may not be able to prevent any negative effects, and it is morally better if we can avoid negative side-effects.

'Certain to be fatal' (5): the full text reads 'When the risk from the activity is realized in the form of a mishap or illness, how likely is it that the consequence will be fatal?' (Slovic, 2000, p. 87). This is again a rather obvious concern; it clearly is morally relevant if the risk of fatality is certain.

Let us pause to see whether and how these considerations might feature in quantitative methods of risk analysis. Considerations (3) and (5) can be captured in the standard formula 'risk = probability x effect', although the formulations are ambiguous concerning whether they refer to the probability, the effect or both. Consideration (2), 'dread', could be incorporated into a cost benefit analysis by assigning costs to public concern. However, it would be an instrumental way of addressing concerns rather than addressing the problem that causes the concern. Furthermore, it does not distinguish between justified and unjustified concerns. Addressing dread in this way might be a form of what I called the 'populist pitfall' in Chapter 2. It is not clear whether or how considerations (1) and (4), related to controllability, can be incorporated in a CBA. However, a common technique in risk analysis is to include barriers to prevent the manifestation of a hazardous chain of events. This can be a way to address the issue of controllability. In the case of the next set of considerations, it is less clear how they can be addressed by conventional approaches to risk analysis.

'Risks and benefits inequitable' (6): most philosophers consider fairness and equity as inherently morally relevant. However, the issue of fairness is not at all recognized in CBA. Many moral philosophers have criticized utilitarianism and consequentialism in general for this shortcoming, and CBA inherits this criticism. CBA could justify the exploitation of one group by another, no matter how small or large each of the groups, as long as the benefits of the one group outweigh the costs of the other group. This is not to say that such considerations may never be made, and they are often unavoidable. But the point is that we cannot simply add and subtract costs and benefits without looking into issues about fair distribution and possible compensations in more detail. Furthermore, certain costs, such as intentionally killing or enslaving innocent people, might be unacceptable under all or most circumstances. To determine when unfair distributions might be unavoidable, we need to go beyond the data provided by a CBA. This requires explicit ethical reflection.

'Catastrophic' (7), full text: 'Is this a risk that kills people one at a time (chronic risk) or a risk that kills large numbers of people at once (catastrophic risk)?' (Slovic, 2000, p. 87, question 7). In absolute numbers, two risks might be identical in terms of the number of victims they cause, but one kills people one by one while the other kills many people at once. Does this mean that these hazards are morally equal or on a par, or might a chronic risk be worse than a catastrophic one, or vice versa? This is a more complicated issue. From a consequentialist perspective, such as in a CBA, we should be morally indifferent about which of these risks materializes. However, imagine two scenarios. One hazard can kill 1,000 people all at once while the other hazard kills 1,000 people distributed over 1,000 years. The second hazard seems more acceptable because there is a possibility that we are able to reduce the lethal power of the hazard in the future. Hence, I think that it is not trivial to treat both hazards equally as consequentialism would demand. From a consequentialist perspective, this is an irrelevant concern even though it seems reasonable to take it into account.

This problem is related to the following considerations. 'Not easily reduced' (9): it is clear that whether we are able to reduce the riskiness of a hazard or not is morally relevant. If the negative effects of a hazard can be easily reduced, it can be a reason to accept the hazard and to try to reduce its effects.

'Risks increasing' (10): This is obviously important since it affects the acceptability of a risk over time, and it might also change a CBA. A CBA should take expected future developments into account. However, this is not a standard requirement for cost-benefit analyses.

This is directly related to 'Threatens future generations' (8). Many philosophers, environmental scholars and environmentally concerned people think we have a moral obligation to leave a world behind that is worth living in for people who will live long after we have died. CBA does not automatically take into account effects that only manifest in a remote future. If it does not do so, it is morally reprehensible; if it does, this gives rise to further methodological problems concerning the predictability of future events. In addition, the possible ethical permissibility of 'discounting', that is, assigning less value to the well-being of future generations (Parfit, 1984), is highly debated in philosophy. I do not have the space to go into the intricate discussions that this involves. However, these methodological and ethical problems indicate the complexity of including these considerations in a CBA.

'Involuntary' (11): voluntariness is a central moral concept that is directly related to the principle of autonomy endorsed by many ethical theories (cf. e.g. Mill, 1985 [1859]). However, consequentialist approaches such as CBA can allow violations of autonomy since they favor actions that maximize outcomes, regardless of how these outcomes are achieved. Hence, while this is a morally important concern, it is at odds with the standard approach in risk assessment.

'Affects me personally' (12): this consideration is not included in the predominant form of consequentialism, namely utilitarianism, as it does not attach different weights to different people. This can be morally right since it reflects the moral value of equality. However, critics of utilitarianism point out that personal considerations are not morally wrong per se. In the context of risks, this is especially important when it comes to individual risks. For example, if a person undergoes a hazardous activity while having been informed of the risks, and is the only one to do so, he or she might be justified in undergoing the activity even if a utilitarian calculus or a CBA says that it is not worth it. This becomes more problematic when it comes to collective risks. But even in this context, it is interesting that some moral philosophers argue that utilitarianism unduly denies any form of self-interest and personal attachments (Williams, 1973). This may be debatable, but it is not unreasonable to at least take these considerations, which a strict form of utilitarianism does not allow, into account. In contrast, a CBA can be designed from the perspective of a company and exclude explicit ethical considerations for the people involved. This means that it can be motivated purely by self-interest, which is even more clearly morally problematic.

Most of the remaining characteristics are related to a lack of knowledge. In general, we can state that a lack of knowledge about a hazard's possible negative effects is directly morally relevant. It might force us to be cautious. But how does a lack of knowledge affect a CBA? CBA incorporates probabilistic knowledge, but if the basis for statistical analysis is insufficient due to a substantial lack of knowledge, a CBA becomes highly problematic, unreliable or even impossible. Hence, a lack of knowledge is of methodological importance for a CBA. However, the fact that there is a lack of knowledge is in itself not an important explicit moral concern in a CBA. This shortcoming becomes clearer if we look at the remaining characteristics in detail.

'Not observable' (13): this means that people might be unaware that an activity they engage in might give rise to negative side effects. Our conventional alarm systems might not work in these cases. It seems reasonable that we should be extra careful with such risks.

'Unknown to those exposed' (14): this is highly relevant for the same reason as the previous item. In addition, both these items might undermine autonomous decision making since autonomous choice requires adequate information ('informed consent' is an important moral principle in medical ethics and human research ethics). While considerations (13) and (14) state that there is a lack of information, which is a morally relevant predicament, the role of the missing information in cost benefit analysis is unclear.

'Effects immediate' (15): it is clear that if a hazard can kill someone 10 years after the exposure instead of immediately, it is still morally relevant. This presents a problem for consequentialism and a CBA: namely, how should risks that only manifest in the future be taken into account, and how far into the future should they be taken into account? A CBA should address this, but it often fails to do so, and if it does, this gives rise to additional methodological issues (cf. my comments concerning characteristics 7–10).

'New (unfamiliar)' (16): newness in itself does not need to be a reason to not do something. However, it might be a reason to be careful and take precautions given that the possible negative consequences of the hazard and the likelihood that they occur are unknown.

'Unknown to science' (17): here we can make the same point as in the previous consideration. A lack of scientific knowledge about risks can be a reason to apply the precautionary principle in some form or another. The lack of scientific evidence about a hazard's potential negative consequences is an important concern and can be a reason for caution and to demand further research before people are allowed to be exposed to that hazard. Of course, there is a huge debate about the precautionary principle, whether it is applicable at all and if so, in what form; but this mainly concerns the implementation of the principle as a concrete methodology.[6] Being cautious can be morally wise and should not be ruled out in advance. Unfortunately, this is exactly what CBA does by not morally problematizing a lack of scientific evidence.

'Many people exposed' (18): the number of people affected obviously makes a difference, and it also plays a role in standard CBA. However, in CBA the exposure of many people to a risk can be justified if the net benefits to a group of people are large enough, even if that group is small. It is doubtful whether this is morally justified (cf. characteristic 6).

I have argued that all the concerns of laypeople that have been revealed by Slovic's empirical studies are indeed legitimate concerns. Laypeople's considerations point to important, well-established ethical considerations and principles such as 'informed consent' and the 'precautionary principle'. Indeed, in the literature on ethical aspects of risk, the same considerations are brought forward as those that play a role in laypeople's risk perceptions studied by Slovic and his colleagues. However, I argued that almost none of these considerations play a role in CBA. This means that CBA misses out on numerous ethically important issues that do figure in laypeople's judgments.

Indeed, many social scientists and philosophers who study risk argue that CBA and the definition of risk as a function of probabilities and unwanted consequences are insufficient to determine whether a risk is acceptable or not. Whether a risk is acceptable is not just a matter of quantitative information about probabilities and consequences, but also involves important ethical considerations such as fairness, equity, autonomy and motives (Krimsky and Golding, 1992; Shrader-Frechette, 1991; Hansson, 2004; Asveld and Roeser, 2009; Roeser et al., 2012). How risks and benefits are distributed within a society (fairness, equality; Hayenhjelm, 2012) is a morally important consideration. Risks that are imposed against people's will are morally questionable (autonomy, Asveld, 2007). Equally, it is morally significant whether a risk arises from intentional actions and negligence, or if it occurs despite responsible conduct (motives; Ross and Athanassoulis, 2012; van de Poel and Nihlén Fahlquist, 2012).

Hence, interestingly, laypeople, psychologists, social scientists and philosophers share many of the same concerns when it comes to the moral acceptability of risk.

3.4 The Ethics and Metaethics of Intuitive Risk Judgments

Let us now delve deeper into the philosophical foundations of ethical aspects of risk by studying the metaethical presuppositions that are implicit in the work of social scientists. The underlying framework of quantitative risk scholars and social scientists who study risk oscillates between logical positivism and social constructivism (cf. Renn, 1998). These approaches link directly to the approaches discussed in Chapter 2: a technocratic and populist approaches typically presume that only mathematical and descriptive data are meaningful (logical positivism), whereas participatory approaches typically presume that there is no independent truth and that decisions about risks can best be seen as a social construction. However, within philosophy, there are alternative approaches to logical positivism and social constructivism. If we expand these alternative approaches to risk, that might lead to other, potentially more plausible theories and approaches. In what follows I will propose that a promising alternative approach is a form of ethical intuitionism. Ethical intuitionism is an ethical theory that combines moral realism, an account of ethical intuitions as a source of moral knowledge and a pluralism of morally relevant features (cf. Roeser, 2011a for a more extensive discussion of ethical intuitionism). I will argue that we can expand this ethical theory to the domain of risk. This can shed important new light on the sociological and psychological literature on risk by providing a novel perspective on the normative relevance of intuitive risk judgments.

Moral Realism and Ethical Aspects of Risk

Slovic's work can be seen as an attempt to provide the basis for including laypeople's judgments in decision procedures about risk. Initially, Slovic seemed to take scientific risk assessment as a standard against which to evaluate laypeople's assessments. The main argument to include laypeople's considerations in risk assessment was procedural or instrumental: in a democracy, the concerns of citizens should be included even if they are mistaken, if only to avoid public unrest. However, in his later work, Slovic emphasized the importance and legitimacy of laypeople's considerations. Rather than taking the judgments of experts and scientific approaches to risk as a standard, he emphasized that all approaches to risk involve normative considerations, such as which effects to take into account and which methods to use to assess them. Slovic equates this with the claim that all risk judgments are inherently subjective (Slovic, 1992, 1999). However, stating that all risk judgments are inherently normative is not the same as stating that all risk judgments are inherently subjective. It is a philosophically controversial claim to understand normativity as a form of subjectivity. Hence, this type of claim should be supported by philosophical arguments, which Slovic does not provide, presumably because these are common ideas in social sciences. However, I do not think that it is wise to make this claim given the project that Slovic is working on, and there are good philosophical reasons for not understanding normativity as a form of subjectivity.

Many social scientists claim that since all risk judgments, even those of experts, include values, all risk judgments are subjective and socially construed. But this does not follow. Slovic (1999) seems to equate objectivity with what is 'out there' and with what is quantitative, whereas all the following notions are grouped under the label 'subjective': 'social construction', 'values', 'assumption-ladenness', 'judgment', 'intuitions', 'subjective assessment', 'qualitative', 'emotional' and 'contextual'. Some of these notions are by definition subjective, or at least not objective, for example, 'subjective assessment' and 'social construction'. However, the other notions are not necessarily subjective. For example, assumption-ladenness might indicate that we should be cautious in adopting a certain view since it might be arbitrary or not well-grounded. But the fact that a position is assumption-laden does not as such make it subjective. Values, judgments, intuitions, qualitative, emotional and contextual are also not necessarily subjective notions. Judgment, intuition and emotion are 'subjective' in the sense that they are bound to persons who have them, but this holds true for all our mental abilities. The question is whether these abilities can help us assess what is really there. This is a philosophically controversial issue; it is far from philosophically obvious whether emotions, judgments and values are subjective projections or if they are forms of objective discernment. According to many contemporary moral philosophers, moral values are not arbitrary or subjective. Moral judgments are truth-apt (moral cognitivism),[7] and they concern objective moral truths (moral realism; cf. for example, McNaughton, 1988; Dancy, 1993, 2004; Shafer-Landau, 2003; Cuneo, 2007; Enoch, 2011; Parfit, 2011).

It is hard to avoid normative relativism when one understands normativity as nothing more than subjective preferences (be it of individuals or of groups). Normative relativism states that there are no universal, objective normative standards. This makes criticism of other normative points of view problematic. Rather than being a liberating approach, as it is seen and endorsed by many social scientists (cf. for example Jaeger et al., 2001), it leaves one normatively empty-handed. Taken literally, normative relativism implies that it does not matter what we do. All normative judgments would then merely be projections on a normatively blank world. This means that we might just as well throw dice or appoint a dictator to determine what to do. Normative or ethical deliberation would only give people the illusion of doing something meaningful; in the end, any answer would do. This seems highly counterintuitive. In normative or ethical deliberation, we feel that much is at stake, and we try to find a correct answer. Alternatively, any person's preference would be morally right.

However, this would be a rather arrogant view and would make criticism of others impossible (cf. Sauer, 2014 for a rebuttal of neo-sentimentalism along these lines). It would lead to plain contradictions as soon as people have different subjective outlooks. Or if one thinks that normativity is constituted by cultural practices, then the practice of a culture could never really be critically scrutinized. Also, in order to form ethical judgments, we would only need to ask, 'what does my group think?' Doing so, though, would exclude the possibility of criticizing the predominant views in our own culture, which would lead to conservatism. In practice, people often do criticize the dominant views of their culture. If normativity and ethics would be relative (either to individuals or to groups), trying to find correct moral answers would be a futile endeavor, and the critical stance inherent to ethical thinking would lose its bite (Roeser, 2005, 2011a).[8] In moral philosophy, these are very common objections to subjectivism and relativism (Wellman, 1963; Rachels, 1999; Moser and Carson, 2001), but they are seldom acknowledged in other academic disciplines.

In direct contrast with normative relativism, the theory of moral realism states that the truth of a moral judgment depends on how the world is and not on how we think the world to be. This means that, just as in general realism, the objects of our knowledge exist independently of our beliefs about them. The main argument for moral realism appeals to the following intuition: if there were no moral truths, there would be no objective standard against which to evaluate a situation. If, for example, morality was merely subjective or constituted by conventionality or 'ways of life' (Mackie, 1977), it would be hard to see how we can say that one way of life can be better than another. However, our moral intuitions often tell us that certain moral practices are wrong. Consider, for example, racism, or the way in which many cultures treat women. Our intuitions are that these practices are *really*, that is, objectively, wrong. However, if morality were merely subjective or constituted by practices, it would follow, by definition, that no practice could be morally better or worse than another. But the whole point of morality seems to be exactly the opposite: moral judgments are about what kinds of actions are right or wrong. This might sound like wishful thinking or circular reasoning, but it should be understood as 'inference to the best explanation'. Moral realists argue that values and qualitative and contextual features are part of the world or part of how the world really is in an evaluative sense; they are not arbitrary or socially construed projections (cf. Shafer-Landau, 2003; Cuneo, 2007; Enoch, 2011 for recent book-length defenses of moral realism).

The fact that people can disagree about moral issues does not imply that ethics is subjective. After all, people also can have different scientific insights, and yet we assume that there are scientific truths that we try to understand. Analogously, moral realists think we can try to understand moral truths. Even if there are moral truths, people can still have different ideas about these truths. In some cases we may be indifferent to people who endorse different moral values, but when it comes to fundamental human rights, for example, we should not be indifferent (cf. Roeser, 2005 for a lengthy discussion of universal moral values, tolerance and relativism).

Another reason to endorse relativism is that objective values do not seem to fit in a scientific worldview. However, this presupposes strong and controversial philosophical and metaphysical assumptions about what could be a correct worldview. The idea that there are objective moral norms and values does not need to be at odds with scientific insights. Science tells us how the world is in a descriptive, empirical sense, but science cannot make claims about how the world is in a metaphysical or normative sense, even though scientific findings could be relevant for metaphysical and normative issues. Metaphysical and normative issues are by definition the domain of philosophy.

Quantitative risk judgments take scientific aspects of the world into account, but judgments about acceptable risk concern normative aspects of the world. These normative aspects are of course related to descriptive (scientific and other factual) aspects of the world, but they cannot be reduced to them. Neither of these aspects of the world can be captured by the other kind of judgment, and neither of these judgments is inferior to the other. They both have their own domain in which they are irreplaceable, and they require different sources of insight: descriptive truths can be discovered by scientific methods and empirical observations; moral truths can be assessed through moral judgment. Moral realism allows for the idea that the additional moral considerations about risk mentioned previously are not subjective. In the words of Kristin Shrader-Frechette:

> it is false to say that hazard assessments can be wholly value free (as many naive positivists claim), and it is equally false to assert (as many cultural relativists do) that any evaluation of risk can be justified. That is, some risk evaluations are more warranted, more objective, than others, although none is wholly value free.
>
> (Shrader-Frechette, 1991, p. 30)

Intuitions and Ethical Aspects of Risk

Let us now turn to the notion of intuitive judgments. As stated at the beginning of this chapter, many empirical scholars see intuitive judgments about risk as highly unreliable. However, there are philosophical arguments that state that judgments and intuitions can be an important source of ethical insight. Ethical intuitionism[9] offers a theoretical framework based on which the considerations of laypeople can be understood as justified, reasonable moral intuitions. This approach can provide us with a theoretical framework for Slovic's normative claims.

According to ethical intuitionists, ethical intuitions are the foundations of more complex moral reasoning. Every kind of reasoning has to be based on basic beliefs or intuitions. Otherwise our reasoning would be circular; or it would lead us to an infinite regress; or we would need to make arbitrary assumptions (cf. Reid, 1969b [1788]; Alston, 1993). Ethical intuitionists draw an analogy with mathematics. In mathematics we start from axioms from which we can build more complex theorems. Ethical intuitionists believe that ethical intuitions function in a similar way as axioms in mathematics in that we cannot argue for them any further, but they can still be taken to be justified (cf. Reid, 1969b [1788]). Ethical intuitionists also invoke the analogy of sense perception. Just as our beliefs about the factual, empirical world are based on our sense perception, our complex ethical beliefs are based on basic moral beliefs or intuitions. Basic moral beliefs are 'self-evident'. This means that they are evident in themselves and not by the provision of further arguments; they are 'non-inferential' (cf. Ewing, 1929). Note that the claim that a belief is 'self-evident' does not mean to suggest that a belief is either immediate in time or that it is infallible (cf. Moore, 1988 [1903]). Just as insights in mathematical axioms may take time, so insights in basic moral considerations may also take time. And just as we can err in our basic mathematical and perceptual beliefs, so we can err in our basic moral beliefs (cf. Reid, 1969b [1788]). In addition, Reid emphasizes that we can also directly understand some moral truths that, as a matter of fact, do allow for further justification (Reid, 1969b [1788]).

Ethical intuitions can be understood as perceptions of moral reality, that is, the moral aspects of the world. Ethical intuitions cannot be reduced to or replaced by other kinds of considerations such as scientific truths, be it from the natural sciences or from the social sciences. Replacing ethics by one of the sciences means replacing normative statements by descriptive statements. However, the whole point of normative thinking is to have a critical attitude toward what is descriptively the case. Maybe it is part of human nature to be cruel, but is cruelty a morally good thing? No. A moral insight like this cannot be replaced by a descriptive way of thinking, since it is always possible to ask whether what is descriptively the case is morally good or right. This is G.E. Moore's so-called 'open question argument'. Trying to replace ethics by a descriptive discipline amounts to what Moore called the 'naturalistic fallacy' (cf. Moore, 1988 [1903]).

Ethical intuitionists argue that we have to take our ethical intuitions at face value. This does not mean that ethical intuitions cannot be mistaken, but they are 'innocent until proven guilty'. Intuitions help us to (morally) assess what is really there, albeit in a fallible way. Even if we cast doubt on certain ethical intuitions, we cannot avoid using other ethical intuitions to do this. Ethical reasoning always involves ethical intuitions—that is, basic moral insights. Such basic moral insights cannot be fully replaced by non-ethical insights or by further arguments, even though these can be relevant.[10]

Applied to risk, we can say that insofar as intuitive judgments are aimed at moral aspects of risk, they can be a *prima facie* (i.e., if not refuted) reliable, irreplaceable source of knowledge.

An Irreducible Plurality of Ethical Aspects of Risk

Furthermore, most ethical intuitionists think that there is an irreducible plurality of morally relevant features.[11] We cannot reduce justice, benevolence, happiness, honesty, gratitude, promise keeping, etc., to each other. All these notions are self-evidently morally relevant; we cannot derive them from more fundamental ethical considerations or principles. This distinguishes intuitionism from Kantianism and utilitarianism, which are monistic theories. Kantianism and utilitarianism both hold that there is one fundamental ethical principle and that all other ethical principles or considerations can be reduced to this. Kantians state that all ethical considerations have to be in accordance with the categorical imperative (Kant, 1964 [1786]). Utilitarians state that the most basic moral principle is that we have to maximize happiness or utility (Sidgwick, 1901 [1874]). Monistic ethical theories cannot account for the fact that sometimes we face conflicting moral demands that we cannot solve by trying to find one 'master principle'. In one situation, consideration A might be more important than consideration B; in another situation, this might be the other way around. This is W.D. Ross's famous account of *prima facie* duties: general moral principles only hold *prima facie*. In concrete cases, one duty can overrule another. There is no pre-established method to judge which duty is more important in which situation, or that provides us with a general serial ordering of moral principles (Ross, 1967 [1930]; also cf. Ewing, 1929). Furthermore, there can be genuine dilemmas in which there are equally good reasons not to do either A or B, but in which there are no other alternatives. Monistic ethical theories such as Kantian ethics or utilitarianism cannot capture the real existence of ethical dilemmas.

Utilitarians reduce all ethical considerations to maximizing outcomes (in terms of factors such as happiness, utility or goodness, depending on the specific version). However, this gives rise to many counter examples, such as involving considerations of autonomy, fairness and equity, or obligations based on what we promised. These are morally important considerations that do not necessarily maximize outcomes. As said previously, CBA, the standard methodology in risk assessment, is modeled on utilitarianism, or more generally, on consequentialism. According to consequentialism, the end justifies the means. The only morally important consideration in consequentialism is which option maximizes outcomes. This means that issues such as the fair distribution of costs and benefits and whether an action is performed voluntarily or not are ignored. Accordingly, consequentialist approaches such as utilitarianism and CBA require us to revise our ethical thinking. Instead, ethical intuitionists take the side of our ethical intuitions. Intuitionists state that there is an irreducible plurality of morally relevant considerations that have to be balanced on a case-by-case basis. On these accounts, context-sensitivity is not a form of subjectivity, *pace* Slovic and other social scientists; rather, the morally relevant circumstances are part of the world and can be objectively assessed. This holds for purely empirical information that can be morally relevant, but it also holds for so-called thick moral concepts that comprise empirical and evaluative aspects.

Jonathan Dancy even goes further than that: he believes that there are no necessary moral principles. He argues that there are only context-specific moral truths, determined by the concrete circumstances in a specific situation. Moral rules should be compared with inductive rules in the empirical sciences; they are merely generalizations that might have a heuristic value, but they are not a guarantee that a certain morally significant factor in most situations will be significant in all situations. In some situations, it might not be important at all, or its significance might even be reversed. For example, honesty might generally be a good thing, but it can be very bad if all it does is hurt somebody. Dancy's opponents respond to this by suggesting invoking more complex moral principles. But Dancy's reply is that we can always think of new situations that undermine a complex rule and that if we are capable of making up candidates for complex general moral rules, why do we not use this ability to make particular moral judgments without ultimate reference to general rules? Apparently we are able to form complex moral judgments anyway (for all this, cf. Dancy, 2004).

Let us look how we can apply these ideas to ethical judgments about risks. As indicated, ethical intuitionists think that we should start out with taking our ethical judgments at face value. In that sense, ethical intuitionism is a common-sense approach. Most people know that it is (generally) wrong to steal, kill, lie, and so on. Moral philosophers might be able to reason more explicitly about ethics, but their intuitive moral judgments are not by definition better than those of people who have never read any scholarly literature on ethics (cf. Reid, 1969b [1788]). Furthermore, moral philosophers are far from being saints who never do things they should not do. Ethical intuitionism is a theory that allows the idea that the moral judgments of laypeople can be justified, even if laypeople are not able to articulate a theoretical framework for their moral views. This can shed some light on laypeople's ethical risk judgments. These judgments can be justified moral concerns, whether or not it is possible to come up with further ethical justifications for them. Maybe the only 'justification' that we can come up with is 'this seems self-evident'. Ethical intuitionism is the only ethical theory that explicitly endorses this possibility. Other ethical theories try to avoid appealing to self-evident beliefs, but in practice, they cannot get around basic normative assumptions (cf. Roeser, 2005).

Let us look again at the risk considerations that Slovic et al. found to play a role in laypeople's risk considerations. I already argued that they all seem reasonable or legitimate. We also saw that whereas some considerations can or need to be substantiated with additional arguments, others simply strike us as self-evident. Most of these considerations are also central in various ethical theories. Furthermore, most of these considerations cannot be reduced to one or a few fundamental principles. For example, equity (6) cannot be reduced to concerns about controllability. All these considerations form an irreducible plurality of morally relevant concerns about risks. Hence, ethical intuitionism provides us with an insightful framework to understand laypeople's intuitive risk considerations as judgments concerning objective moral *prima facie* considerations or morally relevant features.[12] Interestingly, the dominant approach in biomedical ethics is also based on a Rossian approach of *prima facie* duties, that is, the *Principles of Biomedical Ethics* by Beachaump and Childress (2012).

Now the point might be made that the possible contribution that intuitionism can make is not very illuminating. It could be argued that it is a trivial or analytic claim that it is morally relevant if something is catastrophic, lethal or unfair. It might be argued that the appeal to intuitionism and self-evidence does not add anything. However, this would be exactly the kind of response that an intuitionist would like to hear, since it would only confirm that we are obviously dealing with morally relevant features. It would only make it worse for traditional risk analysis that it does not take these considerations into account. We saw that all these considerations are insufficiently addressed, or even not addressed at all, in the standard methodology for risk assessment, namely CBA. While CBA might often be an unavoidable basis for judgments on the moral acceptability of risks, it is surely not always sufficient. The intuitive risk judgments of laypeople provide us with a substantial addition to the standard methodology if we want to achieve more adequate judgments about the moral acceptability of risks.

For example, a fair distribution of risks and benefits is morally good in itself, and not just because it happens to be a situation that someone prefers. Autonomy is an important moral value. The fact that people accept greater risks in the activities they have voluntarily chosen is not so much a sign of irrationality or contingent personal preferences, but reflects the centrality of autonomy to our moral life. An important moral principle is 'ought implies can'. Translated to the context of risk, one could formulate the following moral principle: if possible, try to avoid or minimize potentially harmful activities. However, if there are no available alternatives, one might have no choice but to undertake risky activities. These activities can then be morally justified given the specific circumstances, that is, the fact that there are no reasonable alternatives available. Hence, driving a car might be a risky activity that many people nevertheless undertake because they do not have sufficient public transport available. However, people might reject nuclear energy because of the availability of alternative sources of energy, even though car driving may have a higher mortality rate than the use of nuclear energy. Furthermore, a catastrophic event such as a nuclear meltdown might be unacceptable, even though its probability is low. A one-shot, catastrophic risk can be morally more problematic than a chronic, relatively small risk, even though the respective products of probability and effect might be similar. This is because in the case of a chronic risk, such as traffic risks, there are opportunities to improve outcomes, whereas in the case of a catastrophic risk such as a nuclear meltdown, once it manifests itself, it can prove impossible to stop, and the consequences can be disastrous for generations to come.

CBA oversimplifies the complexity of issues involved in deciding what is an acceptable risk. There is no clear-cut method with which to weigh the different considerations that may play a role in a concrete case in which a decision has to be made about what is an acceptable risk. Some propose designing models that give specific weight to the various factors. However, this seems like an *ad hoc* solution to a more fundamental problem. It is unclear whether how to balance various potentially conflicting ethical considerations can be determined in advance. As argued before, modeling moral trade-offs *requires* explicit moral reflection; it cannot *replace* it. We cannot avoid employing moral judgments in order to make a moral assessment of a situation. However, as argued before, this does not mean that judgments about acceptable risk are subjective.

Furthermore, there can be genuine dilemmas. These are situations in which whatever course of action we choose, we do something wrong. In the case of judgments about acceptable risk, this can easily happen because, as Sunstein (2005) emphasizes, almost all options for action involve risks. Sunstein suggests considering a CBA in such situations. However, a CBA may oversimplify the issues as it disregards the additional considerations previously mentioned. A dilemmatic risk decision would be one where we would have to choose between, for example, an option with an equitable distribution between risks and benefits, and an option with a lower net risk but with a less equal distribution, or where more people are involuntarily affected. Although a CBA might provide us with a clear-cut answer, in such cases it is far from obvious that it would provide an answer that we find morally right. In a genuine dilemma, it might be impossible to find a completely irreprehensible action, but we should at least explicitly consider all morally relevant aspects to make sure our judgment is as well-grounded as possible.

The aforementioned additional aspects of risks are morally relevant considerations that have to be taken seriously in their own right. Interpreting them through the framework of ethical intuitionism helps to reshape the discussion concerning the acceptability of risks: intuitive judgments concerning the acceptability of risks are not subjective; on the contrary, they are essential to a full assessment of the moral acceptability of risks.

3.5 Conclusion

Let us return to the beginning of this chapter where I mentioned that intuitions about risks are seen as very unreliable. We can now conclude that they are indeed unreliable when they concern quantitative aspects of risks, but they are necessary and can be legitimate concerning ethical aspects of risks. Whereas the quantitative part of risk assessment should not be based on intuitions but should be done according to scientific methods, evaluations whether a risk is morally acceptable have to involve ethical intuitions. The philosophical theory of ethical intuitionism can provide us with a framework to understand the considerations of laypeople concerning risks (as identified by Slovic) as legitimate moral concerns. Of course, we could also reject CBA purely on philosophical grounds without taking the detour through the judgments of laypeople. However, Slovic has claimed that laypeople have legitimate concerns about risks, and the detailed discussion of these concerns was meant to substantiate this claim. This discussion showed that the concerns of laypeople are legitimate and cannot be reduced to one simple methodology such as CBA. They include normative and evaluative considerations such as justice, fairness, equity, and autonomy.

This approach comes at a price, which is that it does not give us absolute guidelines as to how to make overall evaluations about the moral acceptability of risks. One hope might be that we can establish a serial ordering of ethical concerns, rated by importance. However, it is doubtful whether this is possible. It might not do justice to each possible case and hence lead to oversimplification again. The intuitionist framework might do more justice to the complexity of the ethical aspects of risks than CBA, but this means that discussions about acceptable risks will have to be made on a case-by-case basis, involving thorough ethical reflection instead of applying a clear-cut methodology. This is what the intuitionist Ross has to say about this:

> Loyalty to the facts is worth more than a symmetrical architectonic or a hastily reached simplicity. If further reflection discovers a perfect logical basis for this or a better classification, so much the better.
>
> (Ross, 1967 [1930], 23)

From the previous discussion of legitimate concerns of laypeople, it is clear that CBA does not provide for a 'perfect logical basis' for moral judgments about risks. This means that we should find different ways of doing risk analysis. According to Ross, judgments about morally complex issues are merely 'fallible opinions' where many good judges might disagree (Ross, 1968 [1939], p. 189). In such cases, the best approach might be to hear many different voices and let various people exchange considerations and arguments. This will not guarantee coming up with infallible solutions, but it might be the best we can do in the face of morally complex issues. In any case, this means that in risk assessment we need more than technological and policy expertise; we also need ethical 'expertise', which can for example be provided by concerned citizens who are willing to contribute to ethics panels and other alternative public decision-making bodies. Hence, understanding risk intuitions through the lens of ethical intuitionism supports approaches to participatory risk assessment as discussed in Chapter 2.

So far so good; this sounds like a potential happy ending: philosophers, social scientists, psychologists and laypeople all agree on how to broaden the conventional approaches to risk. Our story could end here, but things are getting more complicated: more recently, emotions have entered the scene of the psychological and social scientific literature on risk, and emotions seem to mean trouble. The following chapters will investigate whether that is a correct view of risk emotions.

Notes

1. For discussions of and references to the various items: items 1–5 see, e.g., Slovic (2000, p. 21); items 6–9, see, e.g., Sunstein (2005, p. 35); item 10, see, e.g., Tversky and Kahneman (1974).
2. Slovic refers to this ambiguity with the formulation: 'perception of actual or acceptable risk' (Slovic 2000, p. 86). 'Actual risk' denotes the quantitative aspect of risk; 'acceptable risk' denotes the normative aspect of risk.
3. For example, cf. Peterson (2012) for an account of consequentialism that incorporates fairness.
4. A problem with Slovic et al.'s full formulation of characteristic 2 is that it asks the subjects to assess the reactions of others: 'is it [a risk] that people have great dread for?' (cf. Slovic 2000, p. 87). Rating this item highly does not need to say anything about the degree of dread that the subjects might feel. Worse, subjects might be mistaken about the amount of dread that others feel. This is an empirical, quantitative issue. In order to address this issue, it would be more straightforward to measure the subjects' attitudes by asking them directly about their own feeling of dread rather than taking the detour of asking them about the feelings of dread of 'the public'.
5. I will come back to the notion of 'dread' at great length in the chapters to follow, where I will specifically zoom in on the role of emotions in risk judgments.
6. For an overview of the debate about the precautionary principle, cf. Ahteensuu and Sandin (2012).
7. This view is even defended by some expressivists, such as Blackburn (1998) and by constructivists such as Korsgaard (1996a, b).
8. Of course, this short discussion cannot even start to do justice to the complexities in the metaethical debate. There are many subjectivists, expressivists and (neo)sentimentalists who argue that they can avoid normative relativism (e.g., Gibbard 1990). For objections against these arguments, cf., e.g., Huemer (2005), Cuneo (2007), Enoch (2011) and Roeser (2011a).
9. Ethical intuitionism has been developed by, for example, the philosophers Thomas Reid (1969 [1788]), G.E. Moore (1988 [1903]), W.D. Ross (1967 [1930], 1968 [1939]), A.C. Ewing (1929) and more recently, Jonathan Dancy (1993, 2004), Robert Audi (2003) and Michael Huemer (2005). The philosophical approach of ethical intuitionism should not be confused with the highly influential social intuitionist approach by social psychologist Jonathan Haidt (2001). Haidt's approach is more in line with Hume's sentimentalism, as he sees intuitions as subjective gut reactions. Ethical intuitionists instead see intuitions

as rational states that provide us with access to objective moral truths (cf. Roeser 2011a for a discussion of the difference between Haidt's approach and that of the ethical intuitionists). I also discuss the difference between these approaches in more detail in Chapter 5 of this monograph.

10. For a book-length discussion of intuitionism, see Roeser (2011a).
11. With the exception of the utilitarian and intuitionist Henry Sidgwick (1901 [1874]).
12. Ross (1967 [1930]) talks about *prima facie*-duties, but we can also extend his account to other ethical notions, such as values and virtues, all of which I mean to capture with the notion 'consideration'. Note that Shrader-Frechette (1991) also refers to Ross's *prima facie*-duties in this context, but she uses the phrase 'subjective value judgments', which goes against the spirit of Ross's theory and that of all other ethical intuitionists. Far from being subjectivists, all ethical intuitionists, including Ross, are moral realists or objectivists, and all of them have argued extensively against subjectivism and other forms of relativism. This is not merely a verbal matter but a substantial philosophical issue, cf. my discussion of this issue at the beginning of this section.

Part II

Reasonable Risk Emotions

This part provides for the epistemological foundations of a new approach to risk emotions, by drawing on work from risk psychology, risk analysis, psychology and philosophy of emotions, epistemology and metaethics.

4 Risk Emotions

The 'Affect Heuristic', its Biases and Beyond

4.1 Introduction

Recent risk perception research focuses on the role of emotions. As it points to ways in which risk emotions can be misleading, most scholars are reluctant to give emotions an important role in debates on risky technologies. This chapter will analyze these arguments and argue that not all supposed emotional biases are necessarily biases, and even if they are, they are not always necessarily due to emotions. More fundamentally, this chapter will critically examine the framework that underlies most empirical risk perception research, which sees emotions as the opposite of, and categorically distinct from, rationality. The following chapters will then develop and build on an alternative theory of emotions that sees them as a form of *practical* rationality.

Paul Slovic has done pioneering work on the risk perception of laypeople. The risk judgments of laypeople differ substantially from those of experts. However, as discussed in the previous chapter, Slovic has shown that this is not so much due to a *wrong* understanding of risk on the part of laypeople, as to a *different* understanding. Whereas experts define risk as a product of probabilities and undesired outcomes to which they apply cost benefit analysis, laypeople include other considerations in their judgments about risks. These considerations include whether risks and benefits are fairly distributed; whether a risk is taken voluntarily; whether there are alternatives available; and whether a risk might be catastrophic. According to Slovic, these are legitimate concerns (Slovic, 2000). In Chapter 3, I referred to various normative ethical theories to argue that Slovic's claim can be justified on philosophical grounds. I also presented an approach based on the framework of ethical intuitionism that provides the metaethical foundations to justify laypeople's risk intuitions. This approach stays very close to the empirical phenomena but provides them with normative support.

More recently, Slovic has also studied the role that emotion or affect plays in laypeople's risk perception. According to Slovic, laypeople rely on an 'affect heuristic': their affective responses largely determine their judgments about risks. As argued in Chapter 3, Slovic sees the risk perception of laypeople as a source of legitimate concerns. However, in his work on the affect heuristic, Slovic seems to see the views of laypeople as prone to bias and in need of correction by scientific evidence. This threatens to undermine the emancipatory claims that are implicit in his more general work on risk perception. I call this apparent incongruity, examined further on in this chapter, the 'Puzzle of Lay Rationality'. I will argue that this Puzzle emerges as a result of the theoretical framework that Slovic applies to his empirical findings: Dual Process Theory.

In the literature on decision making under uncertainty, affective responses are generally seen as biases but also as heuristics (cf. the heuristics and biases literature, e.g., Gilovich et al., 2002). In contrast, rational judgments are considered to be reasonable and normatively justified. This distinction is part of Dual Process Theory, which states that there are two fundamentally different systems by which we process information and form judgments. System 1 is taken to be spontaneous, intuitive and emotional, and system 2 is taken to be slow, reflective and rational. This is also reflected in the title of Daniel Kahneman's recent bestseller, *Thinking Fast and Slow* (Kahneman, 2011), in which he popularized his groundbreaking and highly influential research.

In this chapter, I will discuss the Dual Process Theory framework in detail and argue that it cannot do full justice to the phenomena involved in laypeople's risk perceptions and emotions. Moral emotions such as sympathy and empathy—as well as more basic emotions such as fear—cross the boundaries between the two systems. They have features that are central to system 1 and features that are central to system 2. Partly because of this, these emotions can provide epistemic justification for moral judgments about risks. Rather than being especially prone to be biases that threaten objectivity and rationality in thinking about acceptable risks, emotions can play an important role in attaining an adequate understanding of the moral acceptability of a hazard. I will provide for a different theoretical framework with which we can interpret Slovic's empirical data in a different light, which leads to claims that are more consistent with Slovic's general work on the risk perception of laypeople. In Chapter 5, I will develop an alternative account of risk emotions. I will argue that affect or emotion is an invaluable source of wisdom, at least when it comes to judgments about the *moral acceptability* of risks. However, in this chapter, I will first take a close look at the dominant approaches to risk emotions by Slovic and others.

4.2 The Affect Heuristic

In recent decades, risk perception research has begun to focus on the role of emotions. Paul Slovic, Melissa Finucane and other empirical scholars have studied the role of emotions, feelings or affect in risk perception (cf. for example Alhakami and Slovic, 1994; Finucane et al., 2000; Loewenstein et al., 2001; Slovic et al., 2002, 2004; and Slovic, 2010a). They have coined the terms 'the affect heuristic' or 'risk as feeling' to describe these perceptions (cf. Finucane, 2012 for a review of the literature; several journals have devoted special issues to this topic: *Risk Management* 2008, no. 3; *The*

Journal of Risk Research 2006, no. 2). It turns out that emotions such as dread or fear significantly influence laypeople's risk perceptions.

Slovic et al. (2004) reviews various studies that point to the important role that affective mechanisms play in decision-making processes. For example, Robert Zajonc has emphasized the 'primacy of affect': non-cognitive, affective responses steer our behavior and judgments (Zajonc, 1980, 1984a, b). Empirical studies have shown that feelings such as dread are the major determinant for laypeople's judgments about risks (Fischhoff et al., 1978; Slovic, 2010a; Sandman, 1989). For laypeople, specific risks are affectively loaded, which influences the way in which they rate risks (Slovic et al., 2002; Finucane et al., 2000). Affect serves, as it were, as a mental shortcut. Slovic et al. (2004) coin this the 'affect heuristic'.

According to Slovic and his colleagues, emotions are an important guide in determining our preferences, but emotions can also be prejudiced and closed to new information:

> the affect heuristic enables us to be rational actors in many important situations. But not in all situations. It works beautifully when our experience enables us to anticipate accurately how we will like the consequences of our decisions. It fails miserably when the consequences turn out to be much different in character than we anticipated.
>
> (Slovic et al., 2002, p. 420)

It is often taken for granted that emotions and feelings are irrational and need to be corrected by reason. We also see this reflected in these empirical studies about emotions in risk perception:

> Because *risk as feeling* tends to overweight frightening consequences, we need to invoke *risk as analysis* to give us perspective on the likelihood of such consequences.
>
> (Slovic et al., 2004, p. 320; italics in original)

Most authors propose to correct 'risk as feeling' by what Slovic et al. call 'risk as analysis', by, for example, using scientific information. Similarly, Loewenstein et al. (2001, p. 271) write:

> the risk as feeling hypothesis posits that . . . emotions often produce behavioral responses that depart from what individuals view as the best course of action.

Sunstein thinks that misguided emotions should be corrected by cost benefit analysis:

> The role of cost-benefit analysis is straightforward here. Just as the Senate was designed to have a "cooling effect" on the passions of the House of Representatives, so cost-benefit analysis might ensure that

> policy is driven not by hysteria and alarm but by a full appreciation of the effects of relevant risks and their control. If the hysteria survives an investigation of consequences, then the hysteria is fully rational, and an immediate and intensive regulatory response is entirely appropriate.
> (Sunstein, 2002, p. 46)

Hence, Sunstein thinks that cost benefit analysis is the ultimate arbiter when it comes to evaluations of policies and concomitant emotions. Most scholars writing on risk emotions think that emotions are highly prone to bias by irrelevant factors. As Slovic et al. (2002) summarize:

> Among the factors that appear to influence risky behaviors by acting on feelings rather than cognitions are background mood (e.g. Johnson and Tversky, 1983, Isen, 1993), the time interval between decisions and their outcomes (Loewenstein, 1987), vividness (Hendrickx et al., 1989), and evolutionary preparedness (Loewenstein et al., 2001).
> (Slovic et al., 2002, p. 415)

Hence, authors who write about risk emotions are in agreement that emotions are a feeble basis for risk judgments, and they need to be corrected by rational methods.

4.3 The Affect Heuristic and Its Biases

In this section, I will discuss in more detail the "blind spots" of risk emotions that various authors have identified. I will examine whether these blind spots are indeed due to emotions, and whether these supposed blind spots really are as blind as the various authors take them to be.

Emotions and Risk Attitudes

Various scholars argue that emotions largely determine one's judgments about risks and benefits. For example, Lowenstein et al. (2001) refers to a study by Eisenberg et al. (1998) that moods determine one's judgments about risks and benefits.

> The researchers found that trait anxiety was strongly and positively correlated with risk aversion, whereas depression was related to a preference for options that did not involve taking an action.
> (Lowenstein et al., 2001, p. 273)

That depressed people prefer options where no action is required fits the (general) profile of people with depression. Similarly, and unsurprisingly, anxious people are risk averse. Hence, an individual's affective traits determine his or her risk attitude. Schwarz (2002) also emphasizes the influence of moods on decision making in general.

It makes sense to consider moods as highly prone to bias since moods are not directed toward anything in particular. Hence, they are not responses to a risky activity or technology, yet they determine our attitude toward it. However, moods should be distinguished from other kinds of affective states. For example, Griffith (1997) and Ben-Ze'ev (2000) distinguish between a large variety of affective states. Emotion scholars typically distinguish non-cognitive moods from cognitive emotions. Hence, that moods are likely to be biases does not necessarily say anything about other affective states.

Indeed, Slovic and colleagues have conducted studies on people's feelings toward specific hazards rather than on general moods and emotional traits. In these studies, they made an interesting observation. They found that risk and benefit are negatively correlated in laypeople's judgments (Slovic et al., 2002, p. 410; Slovic et al., 2004, p. 315). Alhakami and Slovic (1994) point out that this is related to the degree of positive or negative affect associated with a hazard.

> [P]eople base their judgments of an activity or a technology not only on what they *think* about it but also on how they *feel* about it. If their feelings towards an activity are favorable, they are moved toward judging the risks as low and the benefits as high; if their feelings toward it are unfavorable, they tend to judge the opposite—high risk and low benefit. Under this model, affect comes prior to, and directs, judgments of risk and benefit, much as Zajonc proposed.
>
> (Slovic et al., 2004, p. 315; italics in original)

Based on the affect heuristic, a person who has a positive feeling about, say, cellular phones, will rate their benefits higher and risks lower than someone whose affect is less positive. This induces a negative correlation between risk and benefit ratings for cellular phones across participants (Finucane et al., 2000, p. 7). In other words, positive feelings toward a technology let us see it in a rosy light, as useful and safe, and negative feelings toward a technology let us see it as dangerous and with few benefits. However, there is no *prima facie* reason to expect that something that has a low risk has a high benefit and vice versa. Rather, risks and benefits can be expected to be *prima facie* logically and causally independent. Hence, the affect heuristic seems to introduce a bias to our risk perceptions. However, we should keep in mind Slovic's earlier claims that the notion of risk has different connotations for laypeople than for experts. Laypeople also include evaluative considerations in their risk perceptions (cf. Chapter 3). I will come back to this point further on.

Probability Neglect or Availability

Sunstein (2005) argues that emotions are prone to let laypeople neglect probabilities:

> Probability neglect is especially large when people focus on the worst possible case or otherwise are subject to strong emotions. When such

> emotions are at work, people do not give sufficient consideration to the likelihood that the worst case will occur.
>
> (Sunstein, 2005, p. 68)

Related to this, Slovic et al. understand what they call 'availability' as a heuristic that lets us focus on easily imaginable risks, even though they may be minor risks. Slovic et al. (2002, p. 414) argue that imagery is more effective than information about relative frequencies:

> Availability may work not only through *ease* of recall or imaginability, but because remembered and imagined images come tagged with affect . . . The highly publicized causes [of death, SR] appear to be more affectively charged, that is, more sensational, and this may account both for their prominence in the media and their relatively overestimated frequencies.
>
> (Slovic et al., 2002, p. 414)

Slovic et al. say here that 'available', frequently published risks are often more sensational, and thereby more appealing to the imagination and more emotionally charged than risks that get less attention in the media, and this clouds our perception of reality. Slovic et al. review various studies that indicate that emotions dominate probabilistic thinking when what is at stake has a strong appeal to emotions, and that the opposite is the case if what is at stake is less affectively loaded:

> When the quantities or outcomes to which these probabilities apply are affectively pallid, probabilities carry much more weight in judgments and decisions. Just the opposite occurs when the outcomes have precise and strong affective meanings—variations in probability carry too little weight.
>
> (Slovic et al., 2002, p. 410)

Hence, emotions can blind us to quantitative considerations. For example, people who suffer from fear of flying are focused on plane crashes, even though these are extremely rare.

Framing

'Framing' refers to the phenomenon that the way in which information on, for example, risks, is presented largely determines people's evaluations about that information (Tversky and Kahneman, 1974; Slovic, 2000; Gigerenzer, 2002). This is a phenomenon that holds for both laypeople and experts. Tversky and Kahneman (1974) for example let doctors judge if they would recommend a cancer treatment to a patient. One group of doctors received information about the effectiveness of the treatment in terms of the probability of survival while the other group received it in terms of probability of death, while the information was statistically equal. Representation in terms of probability of survival led to significantly more positive evaluations of the treatment than representation in terms of probability of death. In this example, the 'framing' effect is due to emotions, because positive emotions are connected with survival and negative emotions are connected with death.

However, 'framing' or presentation biases are not always related to emotions but can also be related to other possible sources of bounded rationality. Gigerenzer (2002) shows that Bayesian representations of probabilities are more confusing—for laypeople and experts—than representations in natural frequencies. This has nothing to do with emotions but with the fact that Bayesian representations require more sophisticated mathematical insights.

Manipulation

Another blind spot of risk-emotions that Slovic and his colleagues discuss is manipulation. Manipulation is related to framing, but it is broader and presupposes that the sender of the information has the intention to steer the receiver of the information in a certain direction, whereas framing can happen without any such intentions.

According to Slovic et al. (2002), affect can be manipulated and thus be misguiding. For example, background music in movies conveys affect and enhances meaning; models in catalogs smile to convey the positive affect of the products they are selling; people with attractive names are valued more highly; food products carry 'affective tags' such as 'new', 'natural' and so on in order to increase the likelihood of being sold. GMOs are called 'enhanced' by proponents and 'Frankenfood' by opponents (Slovic et al., 2002, pp. 416–417).

However, the question is whether such responses are based on higher-order, cognitive emotions or on more unreflected gut feelings. I will come back to this distinction further on.

Natural Limitations

Another blind spot concerns so-called 'natural limitations' of our understanding of risks. According to Slovic, the experiential system that also comprises affect is subject to inherent biases:

> the affective system seems designed to sensitize us to small changes in our environment (e.g. the difference between 0 and 1 deaths) at the cost of making us less able to appreciate and respond appropriately to larger changes (e.g. the difference between 570 deaths and 670 deaths). Fetherstonhaugh et al. (1997) referred to this insensitivity as *psycho-*

> *physical numbing*. Similar problems arise when the outcomes that we must evaluate change very slowly over time, are remote in time, or are visceral in nature.
>
> (Slovic et al., 2002, p. 418)

Slovic et al. give the example of nicotine addiction: "a condition that young smokers recognize by name as a consequence of smoking but do not understand experientially until they are caught up in it" (Slovic et al., 2002, p. 418). Slovic explains this as follows: "Utility predicted or expected at the time of decision often differs greatly from the quality and intensity of the hedonic experience that actually occurs" (Slovic et al., 2002, p. 419). Slovic takes the examples of smoking and of psychophysical numbing as evidence for the shortcomings of the affect heuristic. However, I would like to emphasize that these examples also indicate the failure of the analytical system: apparently, our abstract knowledge is often not very effective in guiding our thoughts and behavior.

Proportion Dominance

A last blind spot in our thinking about risks that I wish to discuss and that according to Slovic et al. is due to affect is proportion (or probability) dominance:

> Ratings of a gamble's attractiveness were determined much more strongly by the probabilities of winning and losing than by the monetary outcomes. [. . .] We hypothesize that these curious findings can be explained by reference to the notion of affective mapping. According to this view, a probability maps relatively precisely onto the attractiveness scale, because it has an upper and lower bound and people know where a given value falls within that range. In contrast, the mapping of a dollar outcome (e.g. $9) onto the scale is diffuse, reflecting a failure to know whether $9 is good or bad, attractive or unattractive.
>
> (Slovic et al., 2004, p. 317)

This is an interesting observation. However, I am not sure what it says about rationality. It seems only reasonable to be agnostic about assessing the value of a given number if the scale and the upper and lower bounds are unknown. Furthermore, whether this phenomenon really says something about the involvement of affect or emotion is debatable. What is the empirical evidence that evaluations are based on emotions? Maybe the explanation is that Slovic et al. equate ratings of attractiveness with emotional ratings, but whether these are really the same is an open question that should be empirically tested. It is not an analytical claim, and it is philosophically controversial whether evaluative judgments such as attractiveness judgments are made by reason or emotion or both.

To conclude this section: it is clear that there are many blind spots in people's assessment of risks and probabilities, but they are not all as blind as they seem, and they are not all clearly based on emotions.

4.4 Dual Process Theory: Emotions Versus Rationality

The theoretical framework that most scholars who work on risk and emotion endorse is Dual Process Theory (DPT, e.g., Slovic et al., 2002). DPT was developed by Daniel Kahneman and others. According to DPT, there are two distinct systems with which we apprehend reality. System 1 is unconscious, fast, intuitive and emotional, while system 2 is conscious, slow, analytical and rational (Epstein, 1994; Sloman, 1996; Sloman, 2002; Stanovich and West, 2002). DPT sees emotions as irrational, unconscious intuitions and as gut reactions that serve as heuristics in decision making under uncertainty, that are prone to bias and that need to be corrected by rational methods. Furthermore, some DPT scholars argue that rational beliefs are an afterthought to our immediate emotional responses (cf. Zajonc, 1984b; Haidt, 2001).

Psychologists use these labels to distinguish between the two systems: Epstein (1994, p. 711) calls system 1 experiential or emotional and system 2 rational. Slovic et al. (2004) adopt this distinction and state that the experiential system is affect oriented and the rational system reason oriented. Sloman (2002, p. 383) distinguishes between an associative system and a rule-based system. Sloman does not mention affective states of any sort, only intuition (cf. Sloman, 1996, p. 7, table 1). He might view intuition as a form of feeling or emotion, but this is at least a controversial use of terminology for philosophers who also acknowledge the existence of rational intuitions, for example concerning insight into logical or mathematical axioms. The labels for these two systems differ, and their specific versions differ in their details, but defenders of DPT argue that there is sufficient overlap between the various models to justify some consensus that there are indeed two systems of mental processing (e.g. Epstein, 1994, p. 714). The views of several of the authors that Slovic invokes in his work on the affect heuristic are discussed in more detail in what follows.

Epstein

Seymour Epstein has developed an account called cognitive-experiential self-theory (CEST) according to which there are two 'interactive modes of information processing, rational and experiential' (Epstein, 1994, 710). He states that:

> There is no dearth of evidence in everyday life that people apprehend reality in two fundamentally different ways, one variously labeled intuitive, automatic, natural, non-verbal, narrative, and experiential, and the other analytical, deliberative, verbal and rational.
>
> (Epstein, 1994, p. 710)

According to Epstein, the experiential system (system 1) has the following features:

1. holistic
2. affective
3. associative
4. behavior mediated by past experiences
5. encodes reality in images, metaphors and narratives
6. more rapid processing
7. slower to change (changes with repetitive or intense experience)
8. stereotypical thinking
9. context-specific
10. 'experienced passively and preconsciously: we are seized by our emotions'
11. self-evidently valid.

System 2 is characterized by the following features:

1. analytical
2. reason oriented
3. logical
4. behavior mediated by conscious appraisal of events
5. encodes reality in abstract symbols, words and numbers
6. slower processing
7. changes more rapidly (with speed of thought)
8. more highly differentiated
9. cross-context processing
10. 'experienced actively and consciously: we are in control of our thoughts'
11. requires justification via logic and evidence.

(Epstein, 1994, p. 711)

At first sight, these oppositions seem largely plausible. However, at closer inspection, they are problematic. For example, it seems that words (item 5, system 2) are needed for narrativity (item 5, system 1). Furthermore, reliabalists in epistemology maintain that even in logic and sense perception, justification ultimately comes to an end and must rest on foundations (Alston, 1989, 1993). Foundationalists say that all our thinking involves intuitions, that is, non-inferential or self-evident beliefs (cf. Reid, 1996a; Ewing, 1941).[1] This makes the opposition between items 11 on these lists debatable. In addition, many philosophers hold that the axioms of logic (item 3, system 2) are self-evident (item 11, system 1) and grasping them involves reason (item 2, system 2). Intuitionists (cf. chapter 3) would say that moral intuitions are self-evident (item 11, system 1), but they are not necessarily rapid, and most intuitionists would say that they are neither emotional or affective. Far from always being preconscious and passive states (as suggested by item 10, system 1), many human emotions are conscious (item 4, system 2), rational and based on reasons (item 2, system 2). I will discuss this in more detail in sections 4 and 5. Thus, many important ways in which human beings apprehend reality seem to transcend the boundaries of DPT as characterized by Epstein. Hence, although there is some plausibility in the oppositions that Epstein suggests, they might be too crude. There are philosophical approaches that provide for more nuanced views of our mental capacities.

Sloman

Steven A. Sloman distinguishes between an associative and a rule-based system. He emphasizes that this is not the same as the distinction between induction and deduction. The latter distinction refers to different argument types, whereas he is interested in different psychological systems. He argues that the psychological distinction cross-cuts the distinction in argument types (Sloman, 1996, pp. 17, 18). Still, the way Sloman characterizes what in his view are two different psychological systems invites the analogy with the inductive-deductive distinction: 'The associative system encodes and processes statistical regularities of its environment, frequencies and correlations amongst the various features of the world' (Sloman, 2002, pp. 380, 381). He also refers to William James, who calls this empirical thinking (Sloman, 2002, p. 380). Hence, it seems as if the associative system at least has central features in common with induction. It is difficult to understand what Sloman means by the rule-based system because he only mentions some features of it without providing us with a definition. In any case, he says that '[r]ule-based systems are productive in that they can encode an unbounded number of propositions . . . A second principle is that rules are systematic in the sense that their ability to encode certain facts implies an ability to encode others' (Sloman, 2002, p. 381). Here are some features which distinguish the two systems that Sloman has identified: 'The associative system is generally useful for achieving one's goals; the rule-based system is more adept at ensuring that one's conclusions are sanctioned by a normative theory' (Sloman, 2002, p. 382), and '[r]ules provide a firmer basis for justification than do impressions' (Sloman, 1996, p. 15).

Hence, Sloman thinks that system 2, by being rule-based, is a better basis for justification than system 1. However, given Sloman's own account of the two systems, one would think that in the case of empirical knowledge, we need system 1. In order to capture the empirical aspects of the world, we need sense perception (in our daily life) and empirical research data and methods (in scientific research), not just mere abstract rules. In that respect, it is remarkable that Sloman opposes perception versus knowledge, by discussing the Müller-Lyer illusion[2]:

> The Müller-Lyer illusion suggests that perception and knowledge derive from distinct systems. Perception provides one answer (the horizontal

> lines are of unequal size), although knowledge (or a ruler) provides quite a different one—they are equal. The *knowledge* that the two lines are of equal size does little to affect the *perception* that they are not. The conclusion that two independent systems are at work depends critically on the fact that the perception and the knowledge are maintained simultaneously.
>
> (Sloman, 2002, pp. 384, 385; italics in original)

Sloman's proposed opposition between knowledge and perception is problematic, as we can also have perceptual knowledge. Knowledge is a success term that can be ascribed to various sources of belief, amongst which perceptual beliefs, provided that they are true, and justified or warranted.[3] There is currently a debate in epistemology that distinguishes (pre-cognitive) 'seemings' from (cognitive) beliefs, and authors also refer to perceptual illusions, so this would be a more philosophically fine-grained way to characterize the phenomena. However, in Chapter 5, I raise some concerns against 'seemings' accounts. Alternatively, one could say that in the case of the Müller-Lyer illusion, we have a false, direct perceptual belief, versus a correct mediated belief that is based on a ruler. But even in the case of the correct belief, we still need perception to read the measurements on the ruler. The purely rule-based system is inadequate for forms of knowledge that require empirical information. According to some moral philosophers, this is also the case with moral knowledge (e.g., ethical particularism, Dancy, 1993, 2004). I will discuss this in more detail in the next sub-section.

Stanovich and West

Keith E. Stanovich and Richard F. West say the following about DPT.

> System 1 is characterized as automatic, heuristic-based, and relatively undemanding of computational capacity. System 2 conjoins the various characteristics associated with controlled processing. System 2 encompasses the processes of analytic intelligence that have traditionally been studied by information processing theorists trying to uncover the computational components underlying intelligence.
>
> (Stanovich and West, 2002, p. 436)

It is noteworthy that Stanovich and West characterize system 1 as 'relatively undemanding of computational capacity', whereas system 2 is supposed to encompass the 'computational components underlying intelligence'. Intelligence here is mainly associated with computational processing. However, human intelligence possesses alternative resources such as narrative and emotional capacities. In philosophy, this has been argued for by phenomenologists such as Scheler (1948), Plessner (1928) and Dreyfus (1992). While these capacities cannot be captured in the computational terms of system 2, since they are reflective and deliberative, they also do not fit into the more instinctive system 1.

Yet, Stanovich and West seem to see 'the tendency toward a narrative mode of thought' (Stanovich and West, 2002, p. 439) as part of system 1 and as an obstacle to normative rationality, which is provided by system 2. They think that system 1 is evolutionarily primary and 'that it permeates virtually all our thinking' (Stanovich and West, 2002, p. 439). Characteristic for this system is, amongst other things, a high emphasis on context, whereas normative rationality, system 2, aims to abstract from context (Stanovich and West, 2002, p. 439). They argue that:

> If the properties of this system [1, SR] are not to be the dominant factors in our thinking, then they must be overridden by System 2 processes so that the particulars of a given problem are abstracted into canonical representations that are stripped of context.
>
> (Stanovich and West, 2002, p. 439)

This might all be correct in areas where formal reasoning is the most appropriate mode of thinking, but there are many domains of knowledge where formal reasoning is insufficient. This is the case with our knowledge of the material world, for which we need sense perception and concrete, empirical information that takes into account contextual aspects. In addition, it is also the case with ethical knowledge, where we need ethical reflection. Knowledge of the material world and of ethics cannot be achieved by formal reasoning only. Contextualization is not necessarily a vice in either domains of knowledge. It might even be unavoidable, as has been argued by moral contextualists or particularists (cf. McDowell, 1998; Dancy, 1993, 2004).

There are some moral philosophers who insist that ethical reflection involves abstracting from the particular context, the most famous of whom is Immanuel Kant. But present-day Kantians also include contextual features in ethical reflection (e.g., Audi, 2003). Ethical intuitionists have long argued that proper ethical decisions should be made on a case-by-case basis, taking into account the features of specific contexts (for example, Prichard, 1912; Broad, 1951; Ross, 1967, 1968; Ewing, 1929; Dancy, 1993, 2004). The main argument is that ethical reality is so complex that we cannot simply apply general rules. Instead, we have to look at the specific aspects of concrete situations. For example, in general it is good to be honest, and that can even be the case when it might hurt somebody. However, in other cases, it is better to be diplomatic about the truth in order to not hurt somebody, depending on the specific circumstances (cf. Dancy's work for various examples). Our intuitive and emotional capacities are capable of sensitizing us to the relevant contextual features (Roeser, 2011a).

It is remarkable that Stanovich and West state as one of the problems of what they call 'evolutionary rationality' (system 1) as opposed to 'normative rationality' (system 2): 'the tendency to see *design* and patterns in situations that are either undesigned, unpatterned, or random' (Stanovich and West, 2002, p. 438). In the area of logic, this might indeed be a problem for what Stanovich and West call evolutionary rationality (system 1), but in the area of ethics, it is more problematic for philosophers who defend a rationalist position (with rationality understood as part of system 2). Rationalists in ethics think that all ethical decisions can be justified by general principles, which we know by reason. Contextualists deny this. Whereas logic, by its very nature, operates in general, de-contextualized terms, our moral life is the messy, contextual world we live in. We would lose touch with our subject matter if we would try to reason about ethics in purely general terms. A contextualist in ethics would say that the moral world is not necessarily patterned, and that this is exactly the reason that ethical reflection should take contextual features into account. An approach that does not do so will assume patterns that are not really there. Hence in ethics this presents a problem for system 2 rather than for system 1, *pace* Stanovich and West.

Logic is not the ultimate normative standard of rationality in all domains of thinking. Logic might be necessary, but it is definitely not sufficient for ethical reflection. However, Stanovich and West seem to imply that it is so by claiming that system 2—usually equated with formal, computational reasoning—is generally normatively superior to system 1.

The general deductive rules of logic apply in cases that are relevantly similar, but logic is inadequate when determining relevant similarities. Logic is an empty system that tells us which deductive inferences are valid given certain premises, but logic cannot tell us anything about the truth values of the premises themselves. To determine these, other modes of thinking are required. For example, in the case of empirical premises, perception and other forms of empirical insights are needed to determine their truth values. And in the case of ethical premises, we need ethical reflection.

Ethical intuitionists emphasize that we cannot avoid what Prichard calls 'an act of moral thinking' (Prichard, 1912). Moral thinking cannot be replaced by other modes of thinking. In any case, probably no moral philosopher thinks that a purely computational approach will be able to render us ethical insights. Kant's rationalist approach in philosophy distinguishes between 'pure reason' and 'practical reason', where the latter cannot be replaced by logic, for example. Practical reason or 'moral thinking' requires that one *endorses* a moral point of view, which computers are not able to do, and no system of formal logic can supply. In addition, there are philosophers who believe that emotions are essential to our moral thinking, a view I will discuss in more detail further on.

To conclude, Stanovich and West's arguments to consider narrativity, contextualization and related intuitions and emotions as normatively inferior to computational system 2 processing are problematic, particularly so in the moral domain. Furthermore, the fact that narrativity, emotions and contextualization cannot be reduced to instinctive responses cast doubts on whether these capacities can be located in system 1. On the other hand, these capacities do not fit into the computational paradigm of system 2. They seem to belong to either both systems or to neither of them.

4.5 Risk and Emotion: Beyond Dual Process Theory

I have discussed the specific ideas of some of the empirical scholars who work on DPT and who are cited by Slovic and his colleagues seeking to underpin their work on the affect heuristic. While many of the ideas of these scholars are interesting and appealing, each of them gives rise to philosophical concerns. In general, my critique is twofold: first, system 2 is not *generally* normatively more correct than system 1; rather, it depends on the domain of knowledge. System 1 is more appropriate for some domains of knowledge and system 2 is more appropriate for other domains. Second, DPT is too crude in sub-dividing the different ways of apprehending reality in two opposing systems (also cf. Moors, 2014). There are ways of apprehending reality that transcend the boundaries of DPT. In other words, they fit into neither or both systems.

For the purpose of this book, the most important question is where we should place emotions. Should there be one place for all affective states? On the one hand, emotions seem to fit into system 1 because they are affective states. But emotions can also be rational (system 2). They can be spontaneous responses (system 1), but they can also be reflective, deliberative states (system 2). Moral emotions are not necessarily and not paradigmatically biases. In addition, many affective states—especially moods, sentiments, dispositional emotions and character traits—are not quick, but slow. Moral emotions such as sympathy, empathy and compassion can arise rapidly, but they can also emerge slowly and require, or at least allow, scrutiny and reflection.

This leaves room for the following three interpretations of the role of emotions within the framework proposed by DPT.

1. In the case of emotions, system 2 is not normatively superior to system 1, as emotions (system 1) can be normatively superior to reason (system 2).
2. Emotions fall into both systems, and the systems overlap.
3. Emotions fall into neither system.

Based on my previous arguments, we can say that at least in some cases, system 1 is normatively superior to system 2. Furthermore, below I will argue that emotions can have aspects of both systems: separately but also simultaneously. I also argue that emotions do not neatly fit into the categories provided by DPT, which gives rise to the suggestion that we should use a separate category for them (and arguably also for other mental states).

Note that in his book *Thinking Fast and Slow*, Kahneman (2011) frequently argues that we should not take the suggestion of two separate systems too literally, as he means it metaphorically. However, he draws substantial conclusions on the idea of two separate systems, one fast (system 1) and one slow (system 2). This makes it hard to see where he stops using the framework metaphorically and starts using it literally. This makes the framework imprecise at best and misleading at worst, given the far-reaching implications about human rationality that many people draw from Kahneman's work. For example, Sunstein explicitly refers to DPT in order to justify banning emotions from decision making about risks, by arguing that while system 1 might be useful in warning us of immediately life-threatening dangers,

> governments, and people making decisions under circumstances that permit deliberation, can do a lot better.
>
> (Sunstein, 2005, p. 87)

Sunstein here explicitly contrasts system 1 (which is assumed to comprise emotions) with deliberation. In the remainder of the book, I will challenge this opposition by ultimately developing what I call an 'emotional deliberation approach to risk' (Chapter 7). Sloman nuances the claims of DPT by saying the following:

> The point . . . is not that both systems are applied to every problem a person confronts, not that each system has an exclusive problem domain; rather, the forms have overlapping domains that differ depending on the individual reasoner's knowledge, skill, and experience.
>
> (Sloman, 2002, p. 382)

Sloman here seems to leave open the possibility that, depending on the problem domain or a person's skills, system 1 can at least be as appropriate as system 2. But even in this 'liberal' reading, Sloman still takes the existence of the two systems for granted. He just allows some leeway as to their application.

The view that underlies DPT either sees affect as not connected to judgment (cf. Hume, 1975, [1739–1740]) or it sees affect as preceding judgment (Zajonc, 1980, 1984a, b). However, this view does not make it clear where our affective states come from. They could arise from irrelevant conditions, which means that there is no reason to expect that the judgments resulting from such affective states are veridical. This might hold for some affective states, but surely not for all. Many of our moral emotions seem to be appropriate responses to situations in which they arise. One might conclude from this that judgment precedes feeling (e.g., Reid, 1969b). But then the feelings would be a mere add-on to our judgment that would not serve any purpose, except perhaps for moral motivation. However, there are empirical and philosophical accounts that suggest that affective states also play an indispensable epistemological role in our practical and moral lives (cf. for example, Frijda, 1986; Damasio, 1994; de Sousa, 1987; Solomon, 1993; Nussbaum,

2001; Roberts, 2003; for a review of philosophical theories of emotions cf. de Sousa, 2003).

Robert C. Roberts (2003) has proposed an account of emotions as 'concern-based construals' whereby emotions combine the following features: our emotions lead us to be *concerned* about something which we *construe* in a certain way. By 'construing', Roberts means that we see something in a certain way: we see a snake *as* dangerous, we see our partner *as* loveable, etc. By 'concern' he means that we care about the object of the construal. By talking about construals instead of cognitions, as does Martha Nussbaum (2001), for example, Roberts allows for the possibility that our emotions can be at odds with our considered judgments. I might be afraid of flying although I know that it is one of the safest means of transportation. Hence, Robert's account *can* do justice to the intuitions underlying DPT. However, in Roberts' account, emotions are far from being *paradigmatically* prone to biases. To the contrary; Roberts believes that in the normal case, emotions are supported by judgment (Roberts, 2003, p. 106).[4] The fact that his theory allows for irrational emotions does not mean that these are typical phenomena. In a similar vein, the Müller-Lyer illusion does not show that sense perception is always or generally misleading. Emotions can provide us with veridical, justified construals, that is, with evaluative judgments that can sustain deliberation and rational scrutinization, and they paradigmatically do so in virtuous moral agents.

Roberts lists paradigmatic aspects of emotions, which I will draw on to illustrate that emotions do not fit in either of the systems that DPT proposes. This is a summary of Roberts' list (pp 60–64):

1. Emotions are paradigmatically felt.
2. Emotions are often accompanied by physiological changes (the feeling of which is not identical with, but is typically an aspect of the feeling of an emotion).
3. Emotions paradigmatically have objects.
4. The objects of emotions are typically situations that can be told in a story.
5. An emotion type is determined by defining leading concepts (e.g. anger about a culpable offence; fear of a threat etc.).
6. In paradigm cases, the subject believes the propositional content of her emotion.
7. Emotions typically have some non-propositional content.
8. Many emotions are motivational.
9. Emotions can be controllable but also uncontrollable.
10. Emotions come in degrees of intensity.
11. Expression of emotion can intensify and prolong an emotion but it can also cause it to subside.
12. Emotions are praiseworthy and blameworthy.

Features 1, 2, 7, 8 and 9 (uncontrollable) appear to be part of system 1 (in being intuitive, affective responses), whereas features 3, 4, 5, 6, 9 (controllable) and 12 appear to be more closely related to system 2 (in comprising conceptual content and controllability). Some features, such as 10 and 11, do not seem to correspond clearly with either system. In any case, I believe that Roberts' list of paradigmatic features of emotions gives reason to doubt the possibility of classifying emotions in either system, and it is at least far from clear that emotions belong to system 1.

To sum up, Roberts' account of emotions as concern-based construals can do justice to some of the plausible intuitions underlying DPT by allowing that emotions can be at odds with our considered judgments, while still showing that paradigmatically, emotions provide us with important insights and practical rationality. On this account, moral emotions transcend the boundaries that are set by DPT.

I will develop an alternative theory of risk emotions in chapter 5. For now, we can conclude that emotions, and specifically moral emotions such as sympathy, seem to have a hybrid character vis-à-vis the model proposed by DPT. It then becomes unclear what conclusions we should draw about their reliability and trustworthiness. Are they 'heuristics-but-biases', like system 1 operations, or are they reflective, justified concerns like system 2 operations? In the previous sections, I argued that it is far from clear that system 2 is normatively superior to system 1. In addition, a more nuanced theory of emotions, such as that developed by Roberts, casts doubt on the neat distinction between mental states proposed by defenders of DPT.

Let us grant Slovic and other scholars who study risk and emotion that affect is an unreliable source when determining the magnitude of a risk. Scientific methods are much better equipped to quantify the occurrence of particular unwanted effects. But affect might be indispensable in determining acceptable risks, including what counts as unwanted effects.

As discussed in the previous chapter, it is now commonplace in the empirical and philosophical risk literature that risk is not only a quantitative notion but also an evaluative notion (cf. e.g. Fischhoff et al., 1981; Shrader-Frechette, 1991; Krimsky and Golding, 1992). Slovic was a pioneer in providing empirical evidence that laypeople's risk attitudes both comprise a richer understanding of risks and include qualitative or moral concerns. However, surprisingly, these points do not get real attention in the literature on risk and emotion, and emotions are mainly discussed in relation to quantitative issues about risks. A possible alternative hypothesis is that risk-related emotions might be directed at evaluative considerations about risk rather than at quantitative considerations.

This is an alternative hypothesis to Slovic's account of the affect heuristic, and it is this hypothesis that I will examine in the remainder of this book. This hypothesis is based on the following two claims: 1. paradigmatically, moral emotions are affective and cognitive at the same time; and 2. in their assessment of hazards, laypeople form an overall judgment as to the (moral) acceptability of a hazard rather than—often wrongly—weighing up the magnitude of risks and benefits. While Slovic generally emphasizes the latter claim in his work (cf. Slovic, 2000), he does not explicitly consider this claim in his work on the affect heuristic. In the next chapter, I will present an alternative theory of risk emotions that supports the first claim, which in turn allows the interpretation of Slovic's empirical findings in accordance with the second claim. The remainder of the book will develop this alternative approach in more detail and discuss its implications.

4.6 Re-assessing the Affect Heuristic

In the previous sections, I analyzed the framework underlying Slovic's work on the affect heuristic, namely Dual Process Theory. I argued that although DPT can highlight very interesting phenomena, it falls short in covering the diverse aspects of several mental states, specifically of emotions. This has repercussions for the affect heuristic, given its indebtedness to DPT. I will now raise several additional concerns about the way Slovic and his colleagues have established the affect heuristic by examining its building blocks in more detail.

Affect, Feelings and Evaluations

Slovic et al. define affect as follows:

> As used here, 'affect' means the specific quality or 'goodness' or 'badness' (1) experienced as a feeling state (with or without consciousness) and (2) demarcating a positive or negative quality of a stimulus. Affective responses occur rapidly and automatically.
>
> (Slovic et al., 2004, p. 312)

Apparently, affect has three properties, namely being: 1. a feeling state; 2. evaluative; and 3. rapid. There are some ambiguities here. Even if there are states that combine all these three properties, it does not follow that a state that has one of these properties automatically has all the others. Yet this seems to be the underlying assumption. In many articles on the affect heuristic, Slovic uses the words feeling, affect and value interchangeably. For example, both Finucane et al. (2000, p. 2) and Slovic et al. (2002, p. 398) papers contain the following passage:

> According to Zajonc, all perceptions contain some affect. "We do not just see 'A House': We see a *handsome* house, an *ugly* house, or a *pretentious* house'".
>
> ([Zajonc, 1980] p. 154)

Here affect seems to be mainly related to an evaluation. Also cf. Finucane et al. (2000, p. 2, note 1):

> Affect may be viewed as a feeling state that people experience, such as happiness or sadness. It may also be viewed as a quality (e.g. goodness

> or badness) associated with a stimulus. These two conceptions tend to be related. This paper will be concerned with both of these aspects of affect.

Note that this is a more careful formulation: it states that evaluation and feeling tend to be related, whereas in the Zajonc quote they are equated. Sjöberg (2006) too has criticized the ambiguous use of the notion affect in the risk-perception literature. Furthermore, rationalist philosophers would not equate evaluative states with affective states. Similarly, philosophers who think that there is a close link between the two would still distinguish them conceptually (cf. Roeser and Todd, 2014), and they would not necessarily see emotional value judgments as rapid. For example, the appraisal of our loved ones is an emotion that can last for a lifetime. Let us look at the temporal aspect of the affect heuristic in more detail.

Time Pressure

Slovic and colleagues (Finucane et al., 2000) did experiments in which subjects had to rate the risks and benefits of hazards on a seven-point scale (from not at all risky/beneficial to very risky/beneficial). Some subjects were in a 'time pressure condition'—that is, they were given limited time (5.2 seconds) to do their ratings—whereas other subjects were given more time to reflect. The subjects in both groups came up with significantly different ratings. Slovic et al. presuppose that the subjects in the time pressure condition used system 1, affect, while the others used system 2, analytical thinking.

However, first of all, it is not clear that every time pressure decision is based on affect. This assumption only holds if affect is the only mental activity that is triggered under time pressure. However, many rational processes can also be fast, for example automated basic mathematical reasoning such as making simple computations, which do not typically involve emotions. Furthermore, even if the assumption that every time pressure decision is based on affect would be true, we do not yet know which modes of thinking the subjects who were not in the time pressure condition used. Was it reason, or emotion, or both? We would only know if system 1 could be equated with emotions and system 2 with reason, and if time pressure unambiguously distinguishes between system 1 and system 2. In other words, if, as purported, all mental states that arise in a time pressure condition are forms of affect (time pressure implies affect), it still does not follow that affect always arises in a time pressure condition (affect implies time pressure) and that reason always arises in a non-time pressure condition. Slovic does not provide an argument or evidence to make this sharp distinction. This might be the underlying assumption of DPT, but as I argued before, emotions can also have features of system 2, that is, slowness and rationality.

In conversation, Paul Slovic told me that he would distinguish between affect (system 1) and emotions (system 2 or both). With affect he means the pro/con evaluations that we make all the time, often unnoticed. With emotions he means strong, passionate feeling states that we are rarely in. Note that this is at odds with Epstein, who more or less identifies emotions with system 1. But to reiterate, I do not think that this captures the whole range of affective and cognitive responses to reality. For example, moral emotions can be understood as evaluative judgments (similar to how Slovic understands affect), yet they do not always have to be intensely felt, and they can be slow, justified, rational and based on reasoning. It is conceivable that the subjects who were not in the time pressure condition used emotions that are more broadly understood. I will expand on these issues in Chapter 5.

In the previous sections I stated that emotions are not typically system 1 states, and that they can be expected to play an important role in people's affective responses to risks. Going back to Slovic's studies about the affect heuristic, this could mean several things: the subjects in the non-time pressure condition were not necessarily in an unemotional state. They might have experienced reflective, reason-based, justifiable emotions such as fear, sympathy, empathy and compassion, rather than having made purely rational judgments. This would mean that they were also not in pure system 2 states as defined by DPT. But even if the emotional responses of subjects were spontaneous, it does not necessarily mean that they were irrational. We have seen that typical system 1 features can be more appropriate for certain domains of knowledge than system 2 features. However, it is not clear whether the more complex emotions that philosophers are interested in can be elicited in a time pressure condition lasting 5.2 seconds. Hence, it is doubtful that the time pressure condition can filter out all the responses that are affective states of some sort. We know that the subjects in the different situations came up with different ratings, but it is not clear that the responses of the subjects in the non-time pressure condition were not emotional. It might very well be that the non-time pressure subjects used reflective, rational moral emotions. Different experiments such as MRI scanning of brain activities or self-assessment surveys would be needed to determine which mental activities were involved in the non-time pressure condition. It should also be noted that the normative status of the various responses to risk has not yet been addressed.

Affect and Risk 'in the World'

Slovic writes that in the cases that he has examined, affect comes before judgment and that affect often tends to mislead us in these cases, at least when making judgments about the magnitude of a risk (Sunstein, 2005, p. 87 and Loewenstein et al., 2001 make similar claims). Slovic makes this point in various articles, but one formulation is especially interesting:

> That the inverse relationship [between risk and benefit, SR] is generated in people's minds is suggested by the fact that risks and benefits gener-

> ally tend to be positively (if at all) correlated in the world. Activities that bring great benefits may be high or low in risk, but activities that are low in benefit are unlikely to be high in risk (if they were, they would be proscribed).
>
> (Slovic et al., 2004, p. 410)[5]

This is a rather peculiar passage. The remark in parenthesis begs questions: there is no guarantee that past or current risk legislation is appropriate. That current risk legislation might *not* be appropriate is exactly one of the possible outcomes of research about public risk perception and acceptable risk. As I argued above, it seems more reasonable that the magnitude of risks and benefits are independent of each other. More fundamentally, it is surprising that in this passage, Slovic seems to presuppose that there is risk in the real world that is independent of anybody's perception or conception of risk. The question is how this 'real risk' could be determined. In conversation, Paul Slovic has pointed out that this claim about the 'real' relationship between risks and benefits is based on a commonsensical hypothesis. However, this is inconsistent with his usual emphasis on the idea that there is no such thing as 'real risk' 'out there', independently of anybody's definition of what counts as a risk. All forms of risk assessment involve normative presuppositions and contestable assumptions (cf. Slovic, 1999). I have discussed this in Chapter 3 and argued that the claim that risk is a notion with normative aspects does not mean that risk is a subjective notion (also cf. Möller, 2012), but it is noteworthy that Slovic himself adapts his own approach to the objectivity of risk in the context of his research on risk emotions.

Quantitative Versus Evaluative Aspects of Risk

Furthermore, there is an ambiguity in the notion of 'risk': it can refer to the magnitude of a risk but also to the moral acceptability of a risk. It is not clear whether subjects in Slovic's studies were supposed to rate the former or the latter. Based on the information provided in Slovic's articles on these studies, it seems that subjects were not given definitions as to what is meant by risk or benefit. Hence, it is not surprising that subjects invoke various connotations with these notions and have views of risk that diverge from those of experts. Based on Slovic's previous work, we know that this is frequently the case. Hence, it would have been helpful if Slovic would have been more specific in the instructions to his research subjects as to which notion of risk he is interested in, in order to eliminate unnecessary ambiguities.

A problem might be how to distinguish between the magnitude of a risk and its acceptability. The conclusions of Slovic's earlier studies were that whereas scientists are supposedly mainly interested in the magnitude of a risk, they also make normative assumptions. It can be useful to conceptually distinguish between the magnitude and the acceptability of a risk. However, Slovic's work shows that it can be difficult to make this distinction in the case of concrete examples. A given measure of the magnitude of a risk implies normative assumptions as to what is a morally important aspect of risk. However, a possible study might be to provide subjects with information about annual fatalities related to a certain hazardous activity, and then ask them to judge how acceptable they find these hazardous activities. This latter question could also be further split up into 'personally acceptable' and 'socially/morally acceptable'.

Hence, the affect heuristic approach gives rise to certain conceptual and methodological worries that would require further development. However, one could argue that these are minor worries and that the affect heuristic gets substantial credibility from the fact that it is indebted to Dual Process Theory, which is well established. That said, in my discussion of several DPT approaches that form the background of the affect heuristic, I have indicated that these approaches are themselves problematic on two levels: one, that system 2 is normatively superior to system 1; and two, by placing emotions solely in system 1.

4.7 The Affect Heuristic and the 'Puzzle of Lay Rationality'

In the previous sections, I have presented the state of the art in empirical research on risk emotions, namely Slovic's work on the affect heuristic and Dual Process Theory. Both approaches are critical toward emotions. I have indicated that there are some theoretical problems with these two approaches. They moreover also have practical and political implications that give rise to ethical worries.

Although Slovic, Kahneman and other Dual Process Theorists acknowledge the possibility that reason and emotion can interact, they see them in principle as distinct—hence the label Dual Process Theory—and they see analytical approaches to risk and uncertainty as superior. In Chapter 3, we saw that Slovic's approach differs from Kahneman's heuristics and biases-research in the following way: where Kahneman sees people's deviations from rational decision theory as mistakes, Slovic argues for a different concept of risk exemplified in laypeople's risk perceptions. Nevertheless, in his research on the affect heuristic, Slovic invokes Kahneman's and other Dual Process Theorists' framework of classifying emotional biases. Slovic says that emotion and reason can interact and that we should take the emotions of the public seriously since they convey meaning and can show us what we value. Still, he sees analytical methods as the final arbiter in decision making about risks. Other scholars even go so far as to say that emotions should either be excluded from decision making about risk (Sunstein, 2005), or at most, they should be accepted as a given that we should take into account in a democracy (Loewenstein et al., 2001, p. 281) or used instrumentally in order to create acceptance for a technology (De Hollander and Hanemaaijer, 2003). This interpretation of risk-emotions threatens to undermine Slovic's earlier emancipatory claims concerning the risk perceptions of laypeople.

Chapters 2 and 3 reviewed important insights from sociological, psychological and ethical research into risk. The research suggests a consensus on broadening conventional, technocratic approaches to risk decision-making by including stakeholders' views and moral values. However, this consensus is apparently threatened by recent findings about the importance of affect and emotions in decision-making on risks. Hence, it appears that risk emotions constitute the following puzzle: the fact that emotions play an important role in laypeople's risk perceptions threatens to undermine the earlier claims about the broader risk rationality of laypeople. We can formalize this puzzle as follows:

According to the studies reviewed in Chapter 3:

P1: Laypeople include other concerns in their risk perceptions than experts (sections 3.2–3.3).
P2: These other concerns are normatively justified (sections 3.4–3.5).

—

Conclusion [prolay]: Laypeople's concerns are a normatively justified contribution to decision making about risk. (Supports approaches to participatory technology assessment (PTA), cf. Chapter 2).

However, this is the current state-of-the art in the literature on the role of risk emotions reviewed in this chapter:

P3: Emotions play an important role in laypeople's risk perceptions.
P4: Emotions are an unreliable source of decision making about risk.

—

Conclusion [antilay]: Laypeople's concerns are not a normatively justified contribution to decision making about risk. [denial of Conclusion [prolay]]

Thus, there seems to be a direct contradiction between earlier literature (1970s–1990s) that argues for the inclusion of laypeople in risk decision making, and more recent literature on risk emotions that seems to establish that we should be suspicious of including laypeople in risk decision making. This is what I call the 'puzzle of lay rationality'. Note that even scholars who adopt Conclusion [antilay] can still support approaches to PTA, but they do so for either instrumental reasons, to create support for a decision, or for democratic reasons, to include the public because they are citizens who have a right to be heard. But Conclusion [prolay] provides for additional justification to include laypeople's concerns in decision making, because laypeople's concerns provide for important evaluative insights to risk.

This is not a merely theoretical or semantic issue. Rather, it can have radical political implications that are ethically significant. In Chapter 2, I reviewed the standard approaches to risk decision making. I showed that it makes a major difference whether people are excluded from decision making as in technocratic approaches or whether they are merely tolerated or included for instrumental reasons as in populist approaches; or whether they are considered worthy of participation as in participatory approaches. But even in participatory approaches, it makes a significant difference whether people are taken seriously with their various ways of relating to the world, or whether they have to fit in a specific paradigm. The standard approaches to risk emotions indicate that we should be careful with people's emotions and let analytical approaches have the final say. Of course, if there are good reasons to be wary of certain purported ways of approaching the world, we should indeed be careful in allowing them into the decision-making arena. But I have already indicated that not all the arguments against risk emotions are entirely convincing.

In the remainder of this book, I will present an account that sees risk emotions as an important source of insight into ethical aspects of risk, which should be explicitly acknowledged in decision making. In this chapter, I have reviewed the literature that aims to establish P3 and P4. In Chapter 5, I will argue that P4 is a dubious premise. Chapter 5 argues that we have good reasons to hold on to the prolay conclusion, by providing for a different theory of emotions.

4.8 Conclusion

In debates about bounded rationality, the culprit is often by definition 'emotion' without further analysis whether it is indeed emotion that undermines our rationality. Not all spontaneous responses are by definition emotional, and not all emotional responses are spontaneous and arational. Not all biases are emotional, not all emotions are biases and not all supposed biases really are biases. Even when spontaneous responses are emotional, that does not mean that they cannot be based on reasonable concerns. Some responses that initially involved a process of deliberation can become internalized and evoked spontaneously without reflection (cf. Gigerenzer, 2007). Not all intuitive processes are irrational and emotional, such as insight in mathematical axioms, and not all emotions are unconscious and spontaneous, such as the love for our family and friends. There are many moral emotions that have a cognitive content, a narrative structure and allow for reflection and deliberation (Roeser, 2011a). Ethical intuitionists have developed accounts that show that even unreflective and spontaneous moral intuitions can be justified and justifiable (e.g., Reid, 1969b [1788]; Ross, 1968 [1939]; Broad, 1951 [1930]). In Chapter 5, I propose an alternative framework of risk emotions that sees them as a source of practical rationality. In Chapter 6, I propose an alternative approach for correcting emotions, namely by emotions themselves.

Notes

1. Note that the notions 'intuition' and 'self-evident' do not entail infallibility (cf. for example Ewing, 1941 and Audi, 2003).
2. The Müller-Lyer illusion is a perceptual illusion in which two equally long lines appear of different lengths due to arrows at their ends that point inwards and outwards respectively.
3. I do not wish to enter here into the epistemological debate on criteria of knowledge, whether knowledge is justified true belief, justified warranted belief or what have you. It is generally agreed upon by epistemologists that knowledge can be applied to perceptual beliefs that are at least true (plus possibly other criteria).
4. Cf. also Green (1992, p. 36, 37) for an alternative approach concerning what he calls '*anomalous* emotions' (my emphasis): he says that they lack cognitive commitment and the rational properties of belief.
5. In more or less the same formulation, this passage can be found in various other articles by Slovic on the affect heuristic, e.g., Slovic et al. (2004, p. 315).

5 The Philosophy of *Moral* Risk Emotions

Toward a New Paradigm of Risk Emotions

5.1 Introduction

It is a common idea that reason and emotion are opposite faculties. The idea that reason and emotion are diametrically opposed is so deeply ingrained in our cultural and intellectual heritage that it is often taken for granted. Reason is seen as providing us with objective, rational information, while emotion is seen as providing us with basic survival mechanisms but not with reliable knowledge about the outside world. In psychology, the dichotomy between reason and emotion is also present in Dual Process Theory (DPT, cf. Epstein, 1994; Sloman, 1996; Haidt, 2001; Kahneman, 2011 for example). As discussed in the previous chapter, this approach is endorsed by many scholars who study emotional responses to risk (Loewenstein et al., 2001; Slovic et al., 2002, 2004; Sunstein, 2005). Even when scholars emphasize the importance of the role of emotions in judging whether a risk is acceptable, they take these judgments as unreliable and subjective (Finucane et al., 2000; Loewenstein et al., 2001; Slovic, 1999). The danger of these approaches is that emotions and underlying moral concerns may be discarded. This is what I called the 'Puzzle of Lay Rationality' at the end of the previous chapter.

In the previous chapter, I argued that DPT is too indiscriminate in categorizing the different ways through which we apprehend reality. This chapter proposes an alternative approach to risk emotions that is based on recent emotion research. Over the last decades, emotion scholars have challenged the dichotomy between reason and emotion. Many leading philosophers and psychologists who study emotions have developed so-called cognitive theories of emotions. They argue that we need emotions in order to have 'practical rationality', meaning to be rational on a practical level, and in order to make well-founded moral judgments. In this chapter, I will show that this idea can shed new light on the study of risk and emotion. Emotions, and especially moral emotions such as sympathy, but in some cases also fear, can draw our attention to important moral values involved in risk decisions that we might otherwise overlook. We can combine this insight about emotions with what I mentioned in Chapter 3 about moral intuitions about risk. By understanding moral emotions as ethical intuitions, we can say that they track evaluative features of risk. This different understanding of risk emotions can shed new light on the affective responses of laypeople. Laypeople's affective responses to risk can then be seen as possibly legitimate concerns that should be taken seriously in decision making about risks. I will conclude this chapter by arguing that this interpretation of risk emotions can provide a solution to the 'Puzzle of Lay Rationality.'

5.2 The Reason-Emotion Dichotomy in Psychology and Philosophy

The highly influential DPT sees emotions as irrational, unconscious states, which can serve as heuristics in decision making under uncertainty, but which are also highly prone to be biased. DPT reflects the common dichotomy between emotion and reason: emotions and intuitions (part of System 1) are taken to be spontaneous gut reactions that are highly unreliable. Instead, reason and rational, analytical processes (part of System 2) are the ultimate source of objective knowledge, although they come at the price of requiring more effort. According to DPT, System 1-responses are fast heuristics that can serve practical purposes in some contexts, for example in small-scale societies earlier in our evolution. However, in other contexts, for example in our contemporary, complex and technologically evolved world, these responses are highly unreliable and normatively inferior to rational, analytical capacities ('System 2'). Some defenders of DPT acknowledge that the two systems can interact, but they see them as distinct in essence. The DPT framework can give affect an important role (as emphasized by Slovic), but it is usually seen as an argument that we should invoke reason as the final arbiter of our affectively formed judgments (Sunstein invokes DPT in order to be dismissive of emotions).

Related to this, social psychologist Jonathan Haidt (2001) has developed what he calls a 'social intuitionist model'. He claims that our moral judgments are formed by spontaneous, intuitive gut reactions, and our rationality at most works as a post-hoc rationalization, being the 'rational tail' that is wagged by the 'emotional dog.' We are primarily steered by affect in our moral judgments. Haidt goes even further than defenders of DPT in denying rationality an important normative role. According to Haidt, reasoning resembles a lawyer rather than a judge: reasoning means arguing for an emotionally pre-established view rather than finding a true answer.

The opposition between reason and emotion also underlies the two dominant traditions in moral philosophy, namely sentimentalism and rationalism. These approaches are usually linked to the works of the German philosopher Immanuel Kant and the Scottish philosopher David Hume. Both take emotions to be subjective. The sentimentalist Hume (1975 [1748–1752]) believes that ethics is based on emotions ('sentiments'), so he concludes that there cannot be objective moral truths. The rationalist Kant (1964 [1785]) instead believes that ethics is objective, and hence emotions should be banned from moral thought; they can at most play a motivational role.[1] Most moral philosophers believe that we have to choose between the two horns of the Hume-Kant dilemma: either take emotions seriously but forfeit claims to objectivity, or reject emotions as being a threat to objectivity.

However, contemporary theories of emotions enable us to reject this dichotomy as a false dilemma. According to recent developments in the psychology and philosophy of emotions, emotions and rationality are not mutually exclusive, but rather, emotions are a form of practical rationality: they are crucial for moral knowledge, and they involve cognitive aspects. In the following section, I will provide for an alternative account of emotions and intuitions based on a cognitive theory of emotions.

5.3 An Alternative Account of Emotions, Intuitions and Values

Ethical intuitions and emotions are currently hotly debated in metaethics and empirical moral psychology. Psychologists claim to debunk confidence in our ability to make well-grounded moral judgments, as these are supposedly based on irrational, unconscious emotions, intuitions and gut reactions (Haidt, 2001; Greene and Haidt, 2002; Greene, 2003, 2007). Psychologists generally use the notions of intuition, emotion and gut reaction interchangeably. However, as I will argue in this section, emotions and intuitions should not be equated with unreflective gut reactions. Furthermore, the fact that a moral judgment is based on intuitions and emotions does not mean that it cannot involve or sustain reflection.

The aforementioned psychologists interpret ethical intuitions and emotions in a 'non-doxastic' way; in other words, ethical intuitions and emotions are not seen as beliefs, judgments or cognitions. Several contemporary philosophers also argue for non-doxastic accounts of ethical intuitions and emotions (Bedke, 2008; Pelser, 2010; Cullison, 2010; Roberts, 2003). Ethical intuitionists, on the other hand, have traditionally understood intuitions as doxastic and cognitive but non-emotional (Prichard, 1912; Ross, 1930). In this section I argue that a doxastic understanding of ethical intuitions can be combined with a cognitive theory of emotions. This provides for a novel understanding of ethical intuitions *as* emotions, *as* doxastic states. This novel approach can shed light on the role of emotions and intuitions in risk perception, as I will show in section 5.4.

I will first discuss ethical intuitions as understood by intuitionists. I will then move on to show how insights from the theory of emotions can be combined with intuitionism. I will sketch how new insights from emotion research allow us to move beyond the common dichotomy between reason and emotion, and then I will show how these insights can be combined with ethical intuitionism. I will discuss possible objections and alternative, non-doxastic interpretations of moral emotions and intuitions. I will conclude that a doxastic interpretation of ethical intuitions and emotions is a promising alternative to the current dominant non-doxastic accounts.

Ethical Intuitions as Doxastic States

Ethical intuitionism is a moral epistemology that had its heydays during the second half of the 20th century. Intuitionists argued that in order to have moral knowledge, we need ethical intuitions, or basic moral beliefs on which we can ground more complex moral judgments.[2]

In our everyday language, we often use the notion of 'intuition' to refer to spontaneous, unreflective, maybe even unconscious responses, gut reactions or feelings. In everyday language, the notion 'intuition' is sometimes used to refer to an obscure, almost esoteric capacity. In this latter context especially, the notion 'intuition' is often used in a derogatory way, referring to a highly suspect, unreliable capacity. It is therefore not surprising that defending the idea that intuitions could be a meaningful source of knowledge meets strong opposition. Based on this popular understanding of intuitions, moral philosophers of the most diverse leanings have scolded the school of ethical intuitionism. For example, J.L. Mackie said that 'just sitting down and having an ethical intuition is a travesty of actual moral thinking' and John McDowell referred to it as a 'bogus epistemology'. Even several contemporary philosophers who work in the intuitionist tradition, such as Jonathan Dancy, have in the past eschewed the notion 'intuition' (cf. Dancy, 2004, pp. 148–150).[3]

Yet, the notion of intuition has received renewed attention by philosophers.[4] This is partially due to the influential work in psychology and decision theory that I mentioned above. Haidt's approach is based on the ideas of the philosopher David Hume, who argued that ethics is grounded in people's sentiments. Haidt also invokes the ethical intuitionist Ross to explain his approach. However, intuitionism and Humeanism are two opposed meta-ethical theories. Contrary to the way psychologists define intuitions, ethical intuitionists do not equate intuitions with gut reactions, emotions and irrationality.

Most intuitionists do not see intuitions as affectively loaded; rather, they believe that moral intuitions are a product of our rational faculty.[5] Ethical intuitionists, like many philosophers and psychologists, believe that only rational judgments can track objective truths. Since intuitionists believe that ethical intuitions can track objective truths, they believe that ethical intuitions are rational states. Intuitionists do not think that intuitions necessarily pop up spontaneously. Rather, intuitionists use the notion of intuition to refer to a non-inferential judgment. By that they mean a judgment that is not logically derived from premises as in a deductive argument, but rather a judgment that can be 1) a basic belief, such as in sense perception, or concerning a mathematical axiom; and/or 2) an overall (all things considered) judgment in which we make a holistic assessment of various aspects of a situation that requires a context-sensitive evaluation.

Let me briefly address what intuitionists mean by non-inferential beliefs in the forms of 1) basic beliefs and 2) holistic assessments of complex situations.

Concerning 1): intuitionists argue that some of our moral beliefs are what can be called 'basic beliefs.' In the ethical domain, basic beliefs can refer either to general moral principles or to concrete moral judgments.

A basic belief is not a conclusion of a deductive argument, but rather, it can itself be a premise in a deductive argument. An example is a belief concerning a mathematical axiom. Indeed, analogously, some intuitionists refer to ethical axioms (cf. Sidgwick, 1901). Examples that intuitionists give for such basic beliefs are, for example, the general moral principle that 'Similar cases ought to be treated similarly' (Sidgwick, 1901, p. 386, 387). According to intuitionists, it is impossible to derive such a principle from more fundamental premises.

Furthermore, some beliefs can be conclusions in deductive arguments based on underlying premises while they can also be formed directly without a logical inference. An example is a particular moral judgment that we can form directly but which we can also justify by drawing on more basic moral beliefs. For example, Judy can have a direct intuition concerning the wrongness of Pete hitting the rabbit, but her intuition can be further justified by a principle of non-maleficence, that is, the principle not to intentionally harm somebody (cf. Ross, 1930, and Beauchamp and Childress, 2006 who have adapted Ross' approach to bioethics).

Even in the case of 'properly' basic moral beliefs, that is, basic beliefs that cannot be derived from more fundamental ethical principles, we make inferences of some sort. We need judgments about non-moral aspects of a situation to form a moral judgment, but the moral judgment is not a *logical* inference since one cannot derive an 'ought' from an 'is' (cf. Moore, 1903/1988 on what he calls 'the naturalistic fallacy'). This moral judgment can best be understood as an 'intuition' (cf. Roeser, 2011a). Intuitions might be the result of a long process of reflection, but they are not inferentially or deductively based on the preliminary ingredients of reflection (cf. Ewing, 1929; Prichard, 1912). According to intuitionists, empirical judgments can serve as what they call 'preliminaries' to forming moral judgments, but not as premises in the strict logical sense (cf. Prichard, 1912, p. 28; Ross, 1927, p. 127; Ewing, 1929, p. 185 ff.).

Concerning 2), intuitions can be understood as the overall—or holistic—assessment of a complex situation. While this involves beliefs about various aspects of the situation, the overall moral assessment is not an additive process or a matter of drawing a logical conclusion from various premises. Rather, it involves a genuine *moral* judgment. Various intuitionists have emphasized that in concrete situations, different morally relevant considerations have to be balanced. This has to happen on a case-by-case basis, bottom-up, in order to take into account the salient moral aspects that are specific to a concrete situation. This is why many intuitionists have argued that ethical judgments cannot be simply derived from general moral principles (Ewing, 1929; Ross, 1930; Dancy, 2004; also cf. Chapter 3 of this book).[6] This is in contrasts with generalists in ethics, such as Kant, who argue that particular moral judgments can and should be derived from general moral principles. According to Kant, the most fundamental ethical principle is what he calls the 'categorical imperative', which is a deontological principle requiring that moral rules should apply universally, to everyone and without exceptions, and that in our actions we should respect persons and not use them as means only. The generalist idea is also endorsed by the intuitionists Sidgwick and Moore, who defend a consequentialist framework (cf. Roeser, 2011a). However, most intuitionists emphasize that particular moral judgments need to be made bottom-up, on a case-by-case basis, as there can be conflicting moral demands that require a context-specific judgment as to which moral demand is the most important one (cf. my discussion in Chapter 3). Intuitions refer to such context-specific judgments with the notion 'intuition'.

Hence, the way intuitionists understand the notion of intuition differs significantly from the common understanding of that notion and from the understanding by psychologists, decision theorists and other philosophers. So much so that one might wonder why intuitionists used the notion of intuition in the first place. The answer might be simply for lack of a better alternative. Indeed, many intuitionists avoid that notion themselves and try to use different notions. For example, A.C. Ewing said that he was willing to give up the notion in order to 'avoid some scandal' (Ewing, 1929, p. 186). In order to emphasize the non-inferential nature of moral judgments, the intuitionist H.A. Prichard uses the formulation 'an act of moral thinking'. He contrasts this with an act of providing an argument, which he explicitly calls a 'process of non-moral thinking' (Prichard, 1912, p. 29). Instead, an 'act of moral thinking' is an 'immediate (or direct) apprehension' of, for example, an obligation (Prichard, 1912, p. 28). The problem with these formulations is that they are not very illuminating, which might invite the suspicion that intuitionism is indeed a 'bogus-epistemology'.

One might be tempted to say that intuitionists have invited the criticism they received by having invoked such misleading notions. In defense of intuitionists, I would suggest that the idea of non-inferentiality can indeed be well captured by the notion of 'intuition'—it is only unfortunate that this notion carries so many other connotations that are not endorsed by intuitionists. It is important to keep in mind that intuitionists use the notion of 'intuition' in a technical, restricted sense.

The question is whether a non-inferential judgment is a basic notion that we cannot analyze in more detail, as traditional intuitionists seem to think, or whether it actually can and should be further elucidated. Traditional intuitionists are rationalists; they believe that ethical intuitions are cognitive, rational states. Rationalists who are also generalists, such as Kantians, aim to provide for an account of general moral principles as constituted or understood by reason. They argue that particular judgments are derived from such principles, top-down. Reason is the predestined source of general knowledge and logical derivations. However, it is much less clear how a rationalistic approach to particular moral judgments on a bottom-up approach such as the one defended by Ross and most other intuitionists—or a purely particularist approach such as defended by Dancy[7]—would work. In the remainder of this section, I will propose an alternative to the rationalist paradigm endorsed by the traditional intuitionists. Contemporary emotion research in psychology and philosophy offers us the resources to understand intuitions paradigmatically *as* emotions, without equating them with non-doxastic states or irrational gut reactions, as is the case with DPT and Haidt's approach, for example.

Beyond Sentimentalism and Rationalism

Emotions are generally seen as the opposite of reason and rationality. The opposition between reason and emotion is also prominent in metaethics. The usual taxonomy of metaethical theories is based on sentimentalist versus rationalist approaches to ethics. Sentimentalist approaches see values as expressions of our subjective emotions. Rationalists ban subjective emotions from credible ethical reflection and state that objective values are constituted or understood through rationality. We can crudely summarize these positions as follows.

Sentimentalists

P1: Reason is objective.
P2: Emotions are subjective.
P3: Ethics is based on emotions.

—>

Conclusion[sent]: Ethics is subjective.

Rationalists

P1: Reason is objective.
P2: Emotions are subjective.
P3': Ethics is objective (denial of sentimentalist conclusion).

—>

Conclusion[rat]: Ethics is based on reason (denial of sentimentalist P3).

P1 and P2 express a dichotomy between reason and emotion that is shared by sentimentalists and rationalists. Sentimentalists and rationalists then differ on the role they assign to emotions in ethics, and on the ontological status of ethical properties or propositions. P1 and P2 seem to force us to choose between sentimentalism or rationalism.

However, the dichotomy between reason and emotion has been challenged by recent emotion research. Psychologists and philosophers who study emotions argue that emotions are not opposed to, but constitute a specific form of rationality. Emotions are needed for practical rationality. The neuropsychologist Antonio Damasio (1994) has studied people with specific brain defects in the amygdala who do not feel emotions anymore and who have lost their capacity to make concrete moral and practical judgments. Amygdala patients score as high on IQ tests as before their amygdalae were damaged by illness or accident. They also still know in general that one ought not to lie, steal, etc. However, their personality has completely changed. Before the impairment, they were normal, pleasant people, but after their amygdala damage, they turned into rude people who do not know how to behave properly toward others and who cannot make practical and moral decisions in concrete situations. Hence, Damasio's work shows that emotions are necessary to make concrete practical and moral judgments. According to Damasio, emotions are 'somatic markers' with which we perceive morally and practically salient aspects of the world.

This also holds in regard to risk. Damasio and his colleagues developed the so-called Iowa-gambling task: an experiment in which people gamble in a lab setting. People without amygdala defects fall within a normal range of risk seekingness and risk aversion, but amygdala patients have no risk inhibitions. They are willing to take major risks that other people find unacceptable. Apparently, our emotions prevent us from taking outrageous risks and are necessary for making concrete moral judgments.

These ideas are supported by work from other emotion scholars in psychology and philosophy who emphasize that emotions are not contrary to knowledge and cognition. Rather, emotions are needed for cognitions, or emotions are themselves a form of cognition.[8] These are so-called cognitive theories of emotions (e.g., in psychology Frijda, 1986; Lazarus, 1991; in philosophy Solomon, 1993; Nussbaum, 2001; Roberts, 2003; Deonna and Teroni, 2012).

Based on these insights we can say that the dichotomy between rationalism and sentimentalism is based on a false dichotomy between reason and emotion. Applying the insights from emotion research to the debate in metaethics allows for a 'third way' between sentimentalism and rationalism, which I propose as follows.

Based on a Cognitive Theory of Emotions

P1: Reason is objective.
P2': Emotions are objective: emotions are necessary for moral knowledge (Damasio evidence; denial of P2 from sentimentalists and rationalists).
P3': Ethics is objective (denial of sentimentalist conclusion).

—>

Conclusion[cogem]: Ethics is based on emotions [weak claim: epistemological, strong claim: ontological].

The strong ontological claim can be found in response dependence theories according to which emotional responses constitute moral values. Response dependence accounts emphasize that emotions have to be appropriate or fitting (e.g., Gibbard, 1990). However, the question is how this appropriateness or fittingness is to be assessed. If ethics is constituted by emotions, then emotions determine moral values, and it is hard to see how emotions can be assessed by the values they constitute, implying a vicious circularity. In addition, diverging emotions would entail diverging ethics, which would lead to relativism and would undermine the realist claim stated by P3'. In order to be able to assess the appropriateness of emotions, there have to be independent standards. In other words, response dependence accounts are bound to collapse into relativism or realism: either the criteria of fittingness are given by emotions, which means that the criteria are relative or circular, or they are independent of emotions, in which case they are a form of realism after all.[9]

I propose to understand P3' in terms of non-reductive moral realism.[10] On that understanding, an ontological (i.e., constitutive) reading of the relation between emotions and ethics would be too strong, as then ethics could be reduced to emotions, which would again give rise to the problems of relativism and circularity. But the formal argument above allows for a weaker reading of the conclusion, namely as an epistemological claim:

> Conclusion 3[cogem-weak] Ethical *insight* is based on emotion, with emotions understood as a (fallible) source of moral knowledge.

This weaker conclusion is compatible with moral realism. This is the approach I will pursue in the remainder of this section. I will build up the following three claims accordingly.

> Claim 1: Moral emotions are necessary for moral knowledge.
> Claim 2: Moral emotions contain moral judgments.
> Claim 3: Moral intuitions are paradigmatically moral emotions, with both understood as doxastic states.

The claims get stronger as they ascend from 1 to 3. I will discuss these claims one by one, and I will defend them against objections.

Claim 1: Moral Emotions Are Necessary for Moral Knowledge

Cognitive theories of emotions emphasize the importance of emotions when it comes to our appraisal of values. Emotions are intentional states (Goldie, 2000) that draw our attention to what matters (Little, 1995; Blum, 1994). Moral emotions provide us with privileged *focus* on moral aspects of situations. Margaret Little argues that without feelings and emotions, we would not be able to see certain morally relevant features (Little, 1995, p. 127). According to Lawrence Blum, only somebody who cares about certain moral issues can be receptive to the relevant aspects of situations (Blum, 1994). Feelings of sympathy, responsibility and care help us to see what other people's needs might be and that we should help them (the importance of care is also emphasized by feminist ethicists).

Emotions can provide for a deeper *understanding* of the value of a situation. Consider the difference between 'I know that being in state p is bad' in a detached, abstract way, and 'I fully understand how bad it must be to be in state p', where the latter is based on emotional involvement through past experience, sympathy, and empathy. For example, if we hear that somebody has a loved one who is suffering from cancer, we know that this is terrible. However, if we have had a similar experience, we have a much deeper and more thorough understanding of the suffering that these people are going through. We have first-hand experience of what it means to be in such a situation, and we know all the complex emotions and the impact that these events and emotions have on people's lives. However, it is not only through first-hand experience that we attain this capacity of understanding others' emotions and the values at stake. Works of fiction, such as literature and film, can serve as ways to expand our emotional and imaginative capacities and to get closer to someone's first-hand experience, even though we have not literally been in the same situation (cf. Nussbaum, 1992).

There are philosophers who deny claim 1, that moral emotions are necessary in order to have moral knowledge. Most prominently, rationalists (e.g., Kantians) in philosophy argue against this claim. Claim 1 is also challenged by research by social psychologists and other scholars who work within the framework of DPT. Rationalists as well as Dual Process Theorists believe that emotions have dubious epistemological status as compared to rational modes of thinking. The main challenges from the opponents to claim 1 can be summarized as follows: (a.) the evolutionary origin of our moral emotions is taken to undermine their reliability; and (b.) emotions are prone to manipulation by irrelevant factors and hence do not reliably contribute to moral knowledge. I will address these points one by one.

Concerning point (a.): our moral emotions can be explained on evolutionary grounds, which is supposed to undermine claims as to their normative authority. For example, moral emotions such as sympathy for those who are close by can be explained by the development of instincts that were conducive to survival in small-scale societies. The evolutionary basis of moral emotions is seen as a debunking of the legitimacy of moral emotions (Greene, 2003). However, many cognitive systems have evolutionary origins. That a cognitive system has evolutionary origins does not by definition undermine its reliability. Otherwise this would be detrimental to all our faculties. Opponents of claim 1 argue that rational capacities are reflective and in that sense differ from our unreflective emotions and gut reactions, which they consider to be more instinctual. However, emotions can also be a source of reflection and critical deliberation. We can invoke second-order emotions in order to reflect on our primary emotional responses (cf. Lacewing, 2005). We can appeal to sympathy and compassion to broaden a possibly egoistically biased perspective.[11] Emotion scholars point out that there is a broad range of affective states (cf. Griffith, 1997, Ben-Ze'ev, 2000). While some of these states can indeed be seen as unreflective, irrational states, this does not hold for all emotions. For example, higher-order, cognitive emotions have cognitive, reflective aspects (Frijda, 1986; Zagzebski, 2003; Roberts, 2003).

Concerning point (b.): empirical research shows that emotions are fallible and highly susceptible to manipulation by normatively irrelevant factors and thus, it is argued, emotions cannot be a reliable source of moral knowledge (Haidt, 2001; Greene and Haidt, 2002; Greene, 2003, 2007; also cf. recent work of various social psychologists).[12] However, we should note that all cognitive faculties are fallible. Nevertheless, we generally do not conclude from this that they are unreliable unless we are willing to accept a universal skepticism. Indeed, DPT itself provides us with evidence of the fallibility of numerous sources of knowledge (Gilovich et al., 2002; Kahneman, 2011). For example, perceptual illusions show us the fallibility of sense perception. Problems with processing logical or statistical operations show us the fallibility of our rational capacities. Dual Process Theorists generally see the latter examples as instances of intuitive ('System 1', according to DPT) operations. This might hold in cases where people try to address logical puzzles spontaneously, without reflection. However, many people have problems with logical reasoning even if they try to engage their 'slow', analytical capacities ('System 2'). Not everyone is good in logic or math, and this can be the case despite genuine reflective efforts. The fact that even when people try to think analytically they can fail shows the fallibility of rational capacities.

A rationalist might concede this but nevertheless maintain that when it comes to ethical thinking, an unemotional form of practical rationality is more reliable than emotions. Note that this is actually an empirical claim, which means that empirical research is needed to establish it. Rationalists insufficiently acknowledge this. The work by Haidt, Greene, and others shows that moral emotions can mislead us, and I acknowledge this in my concession that moral emotions, and indeed all sources of knowledge, are fallible. Rationality is also fallible in that it can allow us to follow narrow self-interest, leading to suboptimal and morally dubious decisions and behavior, for example so-called 'prisoners' dilemmas' as discussed in rational choice theory. In such cases, moral, other-regarding emotions can serve as a corrective measure (cf. Frank, 1988). In that respect, emotions and rationality seem to be on a par: they are both fallible ways to acquire moral knowledge. Just like other sources of knowledge, they require attention, critical reflection and correction.

But work from neuroscience (Damasio's patients) and psychopathology (sociopaths) supports my claim that moral emotions are *necessary* for moral knowledge. Amygdala patients and sociopaths do often have an intact or even highly functional rationality, but they lack emotions. Studies of these people show that they also lack the capacity to make concrete moral judgments (cf. Nichols, 2004 for an overview of these studies). Indeed, Damasio's amygdala patients are capable of making *general* moral judgments but not particular moral judgments, which requires taking into account concrete circumstances and contextual features.

However, Kantians might argue that these are not genuine *Kantian* general moral judgments. They might object that sociopaths and amygdala patients lack 'Kantian rationality'—in other words, the kind of practical rationality that according to Kantians gives people the ability to make universal moral judgments that are in accordance with the categorical imperative. The Kantian response entails an empirical claim that can be empirically tested.[13] This test can either confirm or refute the assertion that sociopaths and amygdala patients lack Kantian rationality. The results will pose the following dilemma for Kantians.

If, on the one hand, the test shows that sociopaths and amygdala patients have Kantian rationality, this is problematic for Kantians since Kantian rationality appears to be insufficient for moral decision making in particular cases and for moral behavior, as this is what these people have problems with. This would confirm my claim that emotions are necessary for moral decision making, because even if these patients do have Kantian rationality, they lack emotions, and this is what clouds their decision making, as all their other capacities are intact.

If, on the other hand, the test shows that sociopaths and amygdala patients do not have Kantian rationality, this indicates that Kantian rationality actually needs emotions since all the other parts of the rationality of these patients, such as general intelligence (as measured by 'IQ tests'), means-end reasoning and so on are intact. The sociopaths and amygdala patients thus lack emotions and Kantian rationality and are unable to make concrete moral judgments and to behave accordingly. It is more likely that there is a connection between these capacities than that there is a coincidental correlation between these capacities but no causal link. Given current scientific standards, a clear correlation such as this justifies the conclusion that these capacities are causally linked.

Hence, the empirical evidence seems to support claim 1 that emotions are necessary for moral knowledge. Of course, emotions are not infallible, but this is the case with all our cognitive capacities. Many people need glasses in order to see, but nevertheless, seeing is the paradigmatic way to perceive visual aspects of the world. Analogously, emotions can be misguided by, for example, personal interests. However, without emotions, we would not be able to discern moral saliences, as Damasio's amygdala patients and studies with sociopaths show. These studies indicate that emotions are necessary to perceive moral aspects of the world, just as seeing and hearing are necessary to perceive visual and auditory aspects of the world.

Claim 2: Moral Emotions Contain Moral Judgments

Let me now move to my second claim, which is that moral emotions contain moral judgments or beliefs. The dominant approach in emotion research these days is a so-called cognitive theory of emotions, according to which emotions are not contrary to cognition but involve cognitive aspects (philosophers: e.g., de Sousa, 1987; Greenspan, 1988; Solomon, 1993; Blum, 1994; Little, 1995; Stocker and Hegemann, 1996; Goldie, 2000; Ben-Ze'ev, 2000; psychologists: e.g., Frijda, 1986; Lazarus, 1991). Scherer (1984) argues that emotions are complex states that have cognitive, affective, motivational and expressive aspects. In what follows I will propose to understand moral emotions as 'felt value-judgments.'[14] I will argue that emotions are states that can track evaluative features of the world, and they do so with a specific, affective phenomenology.

Some scholars believe that cognitions precede feelings, which together constitute emotions (Reid, 1969b [1788]; Scherer, 1984). Others propose the opposite model, that emotions are constituted by feelings that give rise to cognitions (Zajonc, 1980; Haidt, 2001). There is evidence for both kinds of emotions, but there is a third form of cognitive emotion. Several emotion scholars have argued that paradigmatically, moral emotions have cognitive and affective aspects at the same time, which are disentanglable (Roberts, 2003; Zagzebski, 2003; Roeser, 2011a). For example, prototypical moral emotions such as contempt, indignation, shame and guilt include a moral judgment (the cognitive aspect), but they also involve affective states. Take the emotion of guilt. Experiencing this emotion involves feeling 'pangs of guilt.' Without the 'pangs,' it is not genuine guilt. However, experiencing this emotion also means that the person has the belief that he/she did something wrong. The feeling aspect and the cognitive aspect of emotions go hand-in-hand. These two aspects are inseparable; they are two sides of the same coin. In paradigmatic cases, it is futile to ask whether the affective or the cognitive response comes first. Experiencing indignation means having a judgment and a feeling—it is not the response to an initial, purely cognitive moral judgment. Forming the judgment and having the feeling go hand-in-hand. Based on factual information of a situation, we form a moral judgment that is cognitive and affective at the same time.

Emotions have intentional objects: I am afraid *of something*, I love *somebody*, I am angry *about something*, etc. Furthermore, emotions paradigmatically contain evaluative beliefs about these objects. For example, my fear of snakes contains the belief that snakes are dangerous. My joy at your promotion contains the belief that the promotion is an important and deserved step in your career. My love for my partner contains the belief that he is loveable. However, these emotions comprise more than just a belief. They also have a specific phenomenology. Feeling fear entails for example an increased heart rate and an 'action tendency' (Frijda, 1986) to run away. Being joyful involves a blissful state in which we see something in a cheerful light. Loving one's partner involves a warm feeling of trust and closeness.

Acknowledging that emotions are affective and cognitive at the same time does not deny that cognitive and affective states can exist separately. There are also pure physiological feeling states that do not have a cognitive element, and moods, which are not about specific objects but rather concern a general outlook on life (cf. Ben-Ze'ev, 2000). However, paradigmatic emotions, especially moral emotions such as shame, guilt and resentment, comprise both cognitive and affective aspects. In addition, it is important to note that we do not always feel our emotions, for example the love for our family. To this end, Wollheim (1999) has introduced the distinction between dispositional (potentially feelable) and occurrent (actually felt) emotions.

The affective phenomenology of emotions is not a trivial add-on to a cognitive state that could be equally informative by itself. The affective phenomenology of emotions provides for a *richness of experience* that cannot be substituted by a purely cognitive state. Here we can draw an analogy with sense perception. Sense perception is much richer than purely propositional evidence. In sense perception, we are submerged in the experience of countless details and their interrelations constituting 'organic wholes' (to borrow this notion from intuitionists) that we are unconsciously aware of but that form essential ingredients to our experience. A phenomenological account of sense perception goes beyond providing a list of propositions. Experiencing with one's senses cannot be replaced by reading up on propositional evidence. Another person's meticulous report on the beautiful aspects of the sunset above the sea cannot capture the direct experience of that sunset. Analogously, we can argue that someone who has purely cognitive moral beliefs but does not experience any related affective states misses an important dimension of what it means to have moral knowledge (cf. Little, 1995 and McNaughton, 1988 who draw the analogy with a colorblind person).

We can follow the analogy with color perception very closely to make the conceptual relationship between the affective and cognitive (doxastic, judgmental) aspects of emotions more explicit by highlighting the following components of color perception and moral perception respectively.

Color Vision

1. Our *capacity* to see colors *enables* us to have [capacity, enabling]
2. color *vision*, that is we see a color, which [perception]
3. paradigmatically (unless refuted) comprises a color *belief* [contains belief]
4. and *if* that color belief is *justified and true* (plus possible additional conditions) [if JTB+],[15]
5. then we have (visual) color *knowledge*. [then knowledge]

Moral Emotions

1. Our *capacity* to feel moral emotions *enables* us to have [capacity, enabling]
2. a '*felt value judgment*', that is we feel a moral emotion, which [perception]
3. paradigmatically (unless refuted) also comprises a moral *belief* [contains belief]

4. and *if* that moral belief is *justified and true* (plus possible additional conditions) [if JTB+],
5. then we have emotional moral *knowledge*. [then knowledge]

The schematic representation of color perception and moral perception show that both can be understood in analogous ways. Just as color perceptions (and other forms of visual perception) are unitary states that comprise sensations and beliefs that track visual features, paradigmatically, moral perceptions can be understood as emotions that are unitary states comprising affective aspects and beliefs that track moral features.

Claim 3: Moral Intuitions Are Paradigmatically Moral Emotions, With Both Understood as Doxastic States

I will now turn to how the insights about emotions from the previous subsections can be combined with ethical intuitionism. This involves establishing my third core claim. According to claim 3, ethical intuitions are paradigmatically cognitive moral emotions. The idea is that through emotions we can directly perceive objective moral aspects of the world. They are (fallible) perceptions of moral truths. This approach resembles the moral epistemology proposed by Humeans and other sentimentalists. However, it allows for much stronger metaphysical commitments by being compatible with non-reductive moral realism as advocated by intuitionists. According to this approach, moral emotions are not *projections* on a normatively blank world as sentimentalists hold. Rather, they are the 'window on the world' by which we are *receptive* to the evaluative aspects of the world.

The analysis of moral emotions provided in the previous subsections can be connected to the understanding of moral judgments of ethical intuitionists, namely as direct perceptions of moral salience. The difference is that intuitionists do not acknowledge the affective aspect of moral judgments. Intuitionists (as other rationalists such as Kantians) at most assign emotions a motivating force. However, as the previous discussion has aimed to show, affective states can play an important epistemological role to acquire moral knowledge. A cognitive theory of emotions allows us to combine the Humean idea that emotions are essential for morality with the intuitionist idea that moral judgments are truth-apt. Ethical intuitions can be understood paradigmatically as cognitive moral emotions. This involves a combination of ethical intuitionism with a cognitive theory of emotions. I call this approach 'affectual intuitionism' (in Roeser, 2011a I provide a book-length argument for this approach).

Emotions play an important role in determining whether the suffering of another person is justified or not, and a missing emotion prevents a full-fledged judgment. Paul Slovic has conducted studies that show that people get 'numbed by numbers.' For example, in the case of donating to charities, people are prone to give more money based on the narrative about a single child than on statistical information that presents the full scale of a problem. One would expect a linear relationship between the number of victims and our capacity to care and our willingness to help, but the opposite turns out to be the case (Slovic, 2010b). This work by Slovic shows the limitations of our capacity for compassion, but it also shows the complete failure of our purely rational capacities to respond appropriately to atrocities. This can be overcome by presenting information in a way that appeals to emotions, such as feelings of justice and sympathy, for example by appealing to understandable, gripping narratives (Roeser, 2012). As Nussbaum (2001) has argued, art and narratives can expand our compassion from those who are close by to more distant others. While many emotions are spontaneous responses to what is nearby, sympathetic emotions, for example, can lead us to extend our 'circle of concern', as Nussbaum (2001) phrases it. If we think about the suffering that victims of a disaster might undergo, we usually feel touched and shocked. This realization involves moral emotions. Moral emotions such as sympathy, empathy, compassion, shame and guilt can provide us with access to the moral value of a situation, action or person. Sympathy can broaden our perspective. By feeling with and for another person, I understand that she has the same needs and rights as I do.

As stated at the beginning of this section, intuitions are understood by ethical intuitionists as doxastic states and can be understood in two ways that can also overlap, namely as:

1. a basic belief; and
2. an overall judgment by which we make a holistic assessment of various aspects of a situation that requires a context-sensitive evaluation.

In the remainder of this subsection, I will show that a cognitive theory of emotions allows for a better understanding of these two forms of intuitions.

1) Cognitive Moral Emotions as Basic Moral Beliefs

Moral emotions understood as felt value-judgments can play the role that intuitions and basic moral beliefs play for traditional intuitionists. Moral emotions are not deductive, inferential or strictly argumentative. Rather, through emotions we judge the moral value of a situation in a direct, experiential way. Moral emotions such as sympathy, compassion, shame and guilt provide us with access to the moral value of a situation, action or person. Moral emotions are fundamental moral experiences on which we can base further moral reasoning, which is what Prichard tried to capture with the expression 'an act of moral thinking.' However, the idea that moral intuitions are emotions can give us a much richer understanding of moral intuitions.

Most intuitionists emphasize a bottom-up approach to moral knowledge: we initially make particular moral judgments, based on which we can form more general moral judgments. As mentioned previously, intuitionists do not acknowledge the importance of affective states in this process. However, moral emotions are especially well suited to particular moral judgments. As Zagzebski (2003) argues, moral knowledge starts from concrete, emotional experiences, based on which we form more general moral judgments by 'thinning' out the initial, emotional judgment. According to Zagzebski, general moral judgments are either less intensely felt, or not felt at all.

Just as rational intuitionists have argued about rational intuitions, moral emotions understood as felt value judgments cannot always be scrutinized by inferential, deductive argumentation. However, moral emotions as felt value judgments that consider other perspectives are open to other kinds of reflection such as imagination and empathy. Emotions allow us to critically reflect on other emotions (Lacewing, 2005). Emotions such as sympathy, empathy and compassion let us share the perspectives of others and care for their well-being. Hence, a cognitive theory of emotions provides us with the resources for a richer and more illuminating account of non-deductive moral perception and reflection than rationalist accounts of intuitions. Recall what I said earlier (referring to Ross and Prichard) about the preliminaries of our moral judgments. These preliminaries concern morally relevant empirical information that is needed in order to form a moral judgment about a situation. First, emotions can help to track this information, for example picking up emotional clues such as a sad expression on someone's face, that are relevant for a moral assessment. Second, the moral assessment can involve emotions as well, such as judging that the sad-looking person has been treated unfairly. By not only feeling sorry for the person but also feeling indignation, I pick up the unfairness with which she has been treated. Even though we need certain preliminaries, such as empirical information, for our moral emotions, these preliminaries are not premises in a deductive argument. Rather, we need a genuine moral outlook on the empirical information, which is provided by moral emotions. Hence, moral emotions are not inferential in a strictly deductive sense, just as in the original account of ethical intuitions. However, understanding ethical intuitions as being paradigmatically cognitive moral emotions provides more substance as to the capacities and processes that are involved in forming moral judgments.

2) Cognitive Moral Emotions as Holistic Assessments

When it comes to concrete moral judgments, we often have to assess complex situations that involve a plurality of morally relevant features. Several intuitionists have argued that assessing complex situations cannot be done in an additive way, as the way the features interplay is also important. They think that such holistic judgments are hence best understood as non-inferential or intuitive (Ewing, 1929; Ross, 1930; Broad, 1951).

Based on functional magnetic resonance imaging (fMRI) scans, neuropsychologist and moral philosopher Joshua Greene has argued that different regions of the brain are activated in assessing different sorts of moral dilemmas. People who make impersonal, utilitarian, cost-benefit moral judgments, for example involving hurting somebody who is far away, use rational parts of their brain. People who make deontological, respect-for-persons judgments, for example concerning not hurting somebody who is close by, use emotional parts of their brain (Greene et al., 2001, 2004; Greene, 2003, 2007; Greene and Haidt, 2002). Greene and the moral philosopher Peter Singer (2005) argue that this shows that utilitarian judgments are superior to deontological judgments, as the source of utilitarian judgments is supposedly superior, namely, reason above emotion.

However, the fact that deontological judgments involve emotions does not necessarily undermine their status. Rather, deontological judgments and emotions point to the limitations of utilitarianism and cost-benefit analysis, which is the predominant approach in conventional risk analysis. Making decisions based on utilitarian reasoning might sometimes be inevitable, but there are situations in which respect for persons should be the guiding factor, for example avoiding deliberately sacrificing people to benefit others (Roeser, 2010). Interestingly, there is evidence that sociopaths are more prone to making utilitarian judgments than 'normal' people (Koenigs et al., 2012). As we also saw before in the discussion of amygdala patients, emotions are needed to be sensitive to a variety of context-specific, morally relevant features. This is in line with the intuitionists' idea that morally significantly different situations merit different responses.

Here again, cognitive moral emotions can shed important additional light on our understanding of intuitions. As explored above, concrete emotional moral perceptions or felt value judgments have a unique phenomenology with which we detect complexities that cannot be replaced by a list of propositional statements. It is a rich, felt experience with a unique phenomenology that provides us with direct access to what matters in a situation. This is why emotionally sensitive people, by using their '*Fingerspitzengefühl*', their capacity to make fine-grained distinctions, know how to respond to others in complex situations. Less emotionally sensitive people have to take recourse in general guidelines ('if somebody says they are sad, ask them why' etc.). An emotionally sensitive person has internalized these kinds of responses and does not need to explicitly reflect on them. Such responses are, as it were, second nature to an emotionally sensitive person.

These insights resonate with what intuitionists have said about intuitive moral judgments. For example, C.D. Broad (1951) drew the analogy between intuitive judgments and 'playing games of skill.' These insights also resonate with the description of the *phronimos* by virtue ethicists. Indeed, intuitionism and virtue ethics are closely related in their emphasis on direct moral perception (Ross, 1930; note that Ross also provided a standard English translation of Aristotle's work), and Aristotle is also a source of inspiration for cognitive emotion theorists (e.g., Sherman, 1989). Nevertheless, even though intuitionists have not explicitly connected these aspects of Aristotelian philosophy, doing so would provide for a richer and more convincing account of ethical intuitions as they would be understood *as* emotions.

Problems With Non-Doxastic Accounts of Ethical Intuitions and Emotions

In the previous passages, I have argued that we should understand ethical intuitions *as* emotions, and as 'doxastic' states, that is, as cognitions, beliefs or judgments. However, several contemporary philosophers who study ethical intuitions and emotions deny the idea that these states contain, or are, cognitions, beliefs or judgments; in other words, they deny that intuitions and emotions are doxastic states. In this section I will examine their arguments and voice some concerns.

Some philosophers reject doxastic theories of perception in general and specifically when these theories concern ethical intuitions and emotions. They think that emotions and intuitions are only prerequisites for judgments, proto-judgments or 'seemings' (Audi, 2013; Bedke, 2008; Pelser, 2010; Cullison, 2010; some scholars also use the notions appearance or construal, e.g., Roberts, 2003). Seemings are taken to be non-doxastic states that precede our beliefs or judgments but do not as yet contain a belief or a judgment. Seemings-adherents might endorse claim 1 of my account, but will reject claim 2, and they would formulate my claim 3 in a non-doxastic way.

Matthew Bedke (2008) argues that while self-evidence theories of intuitions are (supposedly) necessarily infallibilist, a problem is that moral intuitions can be mistaken. Hence, he argues, seemings are more able to capture this fallibility as they do not contain beliefs. However, none of the major ethical intuitionists (Reid, Ross, Moore, Ewing) maintained that ethical intuitions are infallible. On the contrary, they all explicitly emphasized that intuitions are fallible. Thomas Reid (1969b [1788]) even stated exactly the same kinds of influences that make our intuitions unreliable as Walter Sinnot-Armstrong in a paper that was critical of intuitionism (Sinnot-Armstrong, 2006; cf. Roeser, 2011a). Nevertheless, intuitionists endorse the notion of self-evidence in order to capture the idea that ethical intuitions are non-inferential (with inferentiality understood in a deductive sense). So intuitionists can maintain that intuitions are self-evident without needing to deny fallibility. Hence, Bedke's argument to endorse a seemings-account for that reason does not hold.

Seeming-scholars argue that our knowledge is based on perceptual states that are initially non-doxastic seemings that turn into beliefs under reflection and full endorsement. Robert Audi defends a non-doxastic account of moral perception, which is not necessarily emotional:

> we may have a *phenomenal sense* of the first [person, SR] wrongdoing the second [person, SR] but *only on reflection form a belief* in which the concept of wrongness [. . .] figures.
>
> (Audi, 2013, p. 54; italics mine)

Similarly, several scholars have defended non-doxastic accounts of emotions. According to Roberts (2003), an emotion is a 'concern-based construal.' He believes that a construal has a propositional structure that is not yet a full-fledged judgment. However, one could also capture Audi and Roberts' ideas through a doxastic account of emotions and moral perception by distinguishing between beliefs and 'considered judgments' that we endorse after reflection. Both the initial belief and the considered judgment could be doxastic, but in the latter case we have more confidence in the truth of our belief.

In a recent article, Dancy (2014) argues that intuitions and emotions are seemings because they are 'presentational':

> Perceptual seemings are conscious, contentful, nonfactive, representational, and presentational. It is in being presentational that they differ from belief or judgment. And qua presentational states, they are baseless, gradable, fundamentally non-voluntary, and compelling, and they tend to make assent seem appropriate.

Dancy first argues that representational states are not necessarily presentational, drawing on the example of blindsight, which is similar to the example of color vision that I discussed before. He then discusses how presentational states are not necessarily beliefs, as in the case of perceptual illusions. Based on this, he concludes that presentational states are seemings rather than beliefs (judgments, doxastic states). However, even if it can be argued that *some* presentational states are not beliefs (in the case of perceptual illusions), it does not follow that *all* presentational states are not beliefs.

Dancy, Roberts and others argue for a non-doxastic account of intuitions and emotions based on the phenomenon of recalcitrant perceptions. For example, in the case of perceptual illusions, even though we know that what we allegedly 'see' is wrong, we still 'see' the perceived object in a way that we know is wrong. The notion of seeming is supposed to capture this. However, I believe that there are various problems with this argument for seemings accounts. How far does one have to go to withhold belief? In other words, when does a seeming turn into a full-blown doxastic state, and how does that happen? Who or what does the judging? Why should we base our epistemology on the exception (perceptual illusions and errors) rather than on the paradigmatic case in which we have veridical perception? I would suggest turning the argument around: rather than seeing seemings as the default that can be turned into full-fledged judgments, the defaults are full-fledged judgments that can be defeated. If defeated, initial beliefs are recalcitrant despite our endorsed belief to the contrary, as in the case of perceptual illusions, these could be called 'seemings.' They are, as it were, 'downgraded' beliefs (cf. Szabó Gendler, 2008 on what she calls 'alief', or Schwitzgebel, 2010 on 'in-between cases of belief'). Hence, the illusions and defective perceptions are the exceptions rather than the upgraded seemings as defended by seeming-scholars.

My argument against a seemings approach is inspired by the intuitionist and Scottish common-sense enlightenment philosopher Thomas Reid. Reid argued that we should take our perceptions at face value and only dismiss and downgrade them in light of contrary evidence rather than working the other way around, as this would eventually lead to skepticism. However, Adam Pelser (2010) has argued that construals are normally followed by beliefs, and that for this reason Reid could have accepted a non-doxastic approach to perception without giving up his anti-skeptical project.

Nevertheless, this gives rise to the Reidian concern as to which faculty we should use to judge whether a perception is reliable. Translated to the context of seemings, this means that the problem is which cognitive or mental resource we should use to judge whether a seeming qualifies as a full-blown cognitive state. As Reid has argued, postponing the ascription of warrant or justification ultimately leads to invoking arbitrary standards, an infinite regress or circularity (Reid, 1969a [1785]; this has also been argued by contemporary 'reliabalists' such as Alston, 1989). The same Reidian argument can be made regarding our moral emotions. Understanding them as seemings raises the question as to which capacities ultimately elevate initial seemings to full-blown cognitive states. Some philosophers argue that although emotions can inform us about moral values, the final verdict should be done by our rational capacities (cf. Brady, 2013). However, as I have argued in the previous passages, the claim that rationality is more reliable than emotions in the context of moral judgments is contentious.

Based on a cognitive theory of emotions, we can understand ethical intuitions *as* emotions, in a doxastic way. Ethical intuitions *as* emotions can then be understood as *prima facie* beliefs or judgments. This leaves open the possibility that they can be defeated, and in that case, they can still be recalcitrant. This means that a doxastic (cognitive) account of intuitions as emotions can capture the phenomena that seemings-accounts address, without falling prey to the worries I have raised against these accounts.

5.4 Re-evaluating Risk Emotions

The recent developments in emotion scholarship and the account I have sketched in the previous section shed new light on the role of emotions in debates about risky technologies. Rather than being seen as opposed to rationality and hence being inherently misleading, emotions should be understood as an invaluable source of wisdom when it comes to assessing the ethical acceptability of risk (also cf. Kahan, 2008). This idea is supported by empirical research: emotions are a form of appraisal of our environment (Lazarus, 1991). They are necessary for avoiding what is bad for us. Emotions are functional in, for example, indicating danger and setting us to act in an appropriate way (Frijda, 1986). Slovic sees laypeople's risk perceptions as legitimate, but he sees emotions as in need of correction by quantitative approaches, which gives rise to the 'Puzzle of Lay Rationality' discussed in Chapter 4. However, my approach to moral emotions and intuitions can shed a different light on Slovic's work on risk perceptions and intuitions.

As I discussed in Chapter 3, ethical intuitionism can provide a justification for Slovic's idea that laypeople's perceptions of risk can be legitimate. My argument in the previous sections that moral emotions are paradigmatically moral intuitions can provide justification to also take people's emotions seriously as important source of insight into evaluative aspects of risk.

As we have seen in Chapter 4, Slovic et al. (2004) and Finucane et al. (2000) seem to equate all immediate responses with affective responses, on the assumption that long-term, reflective responses are not affective. This is based on DPT, according to which system 1 operations are 'rapid, associative, and intuitive', and system 2 operations are 'slow, complex, and often calculative and statistical' (Sunstein, 2005, p. 87). However, my argument in the preceding sections allows long-term reflective responses to be emotional as well, and I argued that purely rational reflection would miss out on important evaluative aspects. Intuitions and emotions transcend the categories of DPT. Intuitions and emotions enable us to have practical rationality; they are sources of knowledge. Hence, they have features of system 1 (e.g., being affective and non-inferential) as well as system 2 (e.g., being reflective and sources of knowledge and justification). While some of our moral intuitions and emotions might be gut reactions, not all of them are. These categories overlap, but they do not coincide. Hence, the fact that a moral insight is based on, or involves, intuitions and emotions does not necessarily mean that it is a gut reaction, and accordingly, that it is irrational.

This can be illustrated by an alternative interpretation of fear or dread, the emotion that Slovic and his colleagues focus on in their work on the affect heuristic. Rather than seeing fear as an unreflective gut reaction, we can understand fear as a perception of what is fearful. Robert Roberts proposes the following defining proposition for fear[16]:

> *X presents an aversive possibility of a significant degree of probability; may X or its aversive consequences be avoided.*
>
> (Roberts, 2003, p. 195; italics in original)

This proposition curiously is more or less identical to the standard account of risk as probability times unwanted effect ('aversive possibility'), although it goes further than that by referring to a 'significant degree of probability' and by stating that X or its consequences should be avoided. According to Roberts, fear is a 'concern based construal.' We try to avoid the object of our fear, we feel an aversion to the object (concern), and we construe it as something worth avoiding. Fear transcends the boundaries of systems 1 and 2. It can be invoked spontaneously, and it can operate very rapidly like a gut reaction (system 1-like), but it also incorporates justifiable factual and evaluative beliefs (system 2-like). Roberts (in criticizing David Hume) emphasizes that fear is paradigmatically based on reasons:

> If someone fears a slippery sidewalk, it makes perfectly good sense to specify his reason(s) for fearing it: He needs to traverse it, his shoes do

> not have good traction on ice, he is unskilled at remaining upright on slippery surfaces—in short, the conditions are such that the slippery sidewalk may well occasion an injurious fall. Apart from some such reasons, it may be as hard to see why a slippery sidewalk would be an object of fear as it is to see why, apart from reasons, a self would be an object of pride.
>
> (Roberts, 2003, p. 194)

Note that Roberts not only cites reasons for a specific fear, he also claims that without these reasons, we could not see how something could be an object of fear. If this holds for fear of slippery sidewalks, surely this holds even more for fear of nuclear power plants, pesticides, explosives and chemical plants, to mention just a few of the hazardous objects that subjects had to rate in Slovic et al.'s studies on the affect heuristic (cf. for example Finucane et al., 2000, p. 7, exhibit 3). We are able to cite justifiable reasons why people would fear hazards such as the destruction of our environment and the possible deaths of human beings. This involves factual beliefs about the possible consequences and evaluative beliefs about the desirability of these consequences. This is what Harvey Green says about fear:

> Imaginative fears, for example, are secondary to ordinary fears involving beliefs about dangers, not just in being less common, but in a more basic sense. Our emotional sensitivity to imaginative representations of danger is explained by our sensitivity to beliefs about danger, for it is in those cases that our emotional representations have the adaptive value which accounts for their evolutionary origin.
>
> (Green, 1992, p. 38)

Here Green makes some interesting remarks: one, genuine fear involves beliefs about dangers; two, it is evolutionary adaptive; three, irrational fears such as in imaginative representations are anomalous and derivative of genuine fear. The latter claim resonates with my previous argument concerning perceptual illusions as the deviation of the default. Fear has adaptive value; it can guide us away from destructive situations. A being without fear would probably not survive for very long. To sum up, fear can be a justified, reasonable concern (system 2-ish) rather than merely a blind impulse (which would be the case if it would be a pure system 1-state).

Furthermore, there is an ambiguity in the kinds of surveys that the research subjects in Slovic's study took. They often had to rate the extent to which they found the current level of risk of a specific hazard acceptable or not. This is ambiguous because the distinction was not made between whether the subjects found the risk acceptable for themselves or for society. They may fear or dread being the victim of the manifestation of a hazard, but they could equally be concerned about the well-being of other people. The latter kinds of concerns are explicitly involved in other-regarding emotions such as sympathy, empathy and compassion.

Many emotions are spontaneous responses to what is nearby, but sympathetic emotions are emotions that allow us to take on a broader, reflective perspective. If we think about the suffering that victims of a disaster undergo, we usually feel touched and shocked. This realization involves emotions. These emotions are reflective, justifiable and based on reasons (unlike system 1), and they can play an important role in our moral assessment of risks. Through these emotions, we can see how morally unacceptable a certain hazard might be. Purely rational reflection might not suffice, as we need the imaginary power of emotions to envisage future scenarios, to take part in other people's perspectives and to evaluate their destinies. Hence, such moral emotions neither fit neatly into system 1 or into system 2. In the next section, I will discuss explicitly other-regarding moral emotions such as sympathy, empathy and compassion. They do not figure explicitly in most of Slovic's studies (except for his more recent work on psychophysical numbing and genocide), but they can shed important light on the study of moral emotions concerning risk.

5.5 Risk Emotions and Moral Considerations

Let us look again at several of the additional considerations concerning acceptable risks that I mentioned in Chapter 3, and that play a role in lay people's risk perception. I will focus on the following considerations: (1) whether a risk is taken voluntarily; (2) the distribution of risks and benefits in a population; (3) the available alternatives to a technology; and (4) whether a higher probability of a small effect might be more acceptable than a small probability of a large effect. In Chapter 3, I argued that these considerations should be seen as reasonable moral considerations that can make a difference as to whether a risk is morally acceptable or not. I will now discuss the extent to which emotions can play a role in assessing these considerations.

(1) Voluntariness

If people are forced against their will to do something that they consider dangerous, it might lead to feelings of anger and frustration. However, these reactions are intelligible: a *prima facie* injustice has been committed. Only if these people can be persuaded that there are good overriding reasons to undergo such risks will their resentment cease. In contrast, if no good explanation is given, they will presumably remain upset. This seems legitimate as one of their fundamental rights has been harmed. Presumably we would find somebody irrational if they would *not* get upset by such an assault on their fundamental rights. If somebody would say, 'I know, they are violating my autonomy by building this chemical factory in my neighborhood without informing me or asking my consent, and without being able to show me the general advantages of it, and I think that it is not fair, but I don't care', we might question their rationality. Somebody who makes a moral judgment should have a fitting emotion as well. Conversely, if somebody does not feel outrage when his or her autonomy is violated, he or she might not be able to fully grasp the injustice that has been done to him or her.

(2) Distribution of Risks and Benefits

Here we can give an argument similar to the one above. A fair distribution of risks and benefits is morally preferable to an unfair distribution. It is reasonable that people feel outrage if they undergo the risks of a certain technology without being able to benefit from it, whereas somebody else may get all the benefits without undergoing the risks. Imagine somebody who lives in a poor neighborhood in which a polluting factory is built, while the director of that factory lives in a wealthy neighborhood at a safe distance from the factory. It is reasonable for the person in the poor neighborhood to feel indignation or even outrage at this situation.

In both cases—the violation of autonomy and the unfair distribution of costs and benefits—it is morally reasonable if the victim of the injustice feels outrage at that injustice. In addition, we would also expect that other people find such injustices morally reprehensible even though they are not themselves the victims. We would even expect that someone who imposes this injustice on another person should be forced to reassess his or her action by caring about the circumstances of the other. If this person does not have feelings of sympathy, we would call him or her hard-hearted and egoistic. Hence, emotions do not only help to assess one's own situation but also that of others, and they can help one see that one's own actions can impose an injustice on others.

(3) Available Alternatives

An important moral principle is 'ought implies can.' Translated to the context of risk, one can formulate the following moral principle: if possible, try to avoid or minimize potentially harmful activities. However, if there are no available alternatives, one might have no choice but to undertake risky activities. In this case, it can be justified.

Hence, driving a car might be a risky activity that many people nevertheless might undertake in the face of a lack of public transport. However, the same people might reject nuclear energy because there are alternative sources of energy even though car driving might have a higher mortality rate than the production of nuclear energy. A person might live in a small village or in a country without a well-organized public transportation system, or she and her partner might have busy jobs while their children also need to be taken to school. In these cases, a car might be the only acceptable means of transportation, even though people might fear the risks of driving a car. However, in the case of nuclear energy, there are various alternative sources of energy that are not as yet fully exploited, such as solar energy and wind energy. The main argument of proponents of nuclear energy is that, as long as everything goes well, nuclear energy is safe and clean and that it is cheaper than alternative sources of energy. However, if alternative sources of energy would be exploited at a larger scale, they might become cheaper in the long run. Nuclear energy is clean and safe as long as there is no accident, but there is still the currently unsolved problem of nuclear waste. Furthermore, while the probability of an accident might be small, there is still always the possibility that an accident may happen and then the consequences would be dramatic. Hence, it is not irrational that people prefer alternative sources of energy and that they have a negative emotional attitude, such as fear, toward nuclear energy.[17]

(4) High or Low Probability or Effect

Let us stay with the same example, driving a car versus nuclear energy. Even if the probability of a car accident might be relatively high, the possible bad effects are rather limited. They range from car damage through slight injury to major injury or even death. Even though major injuries or the death of a few people are shocking, these are nothing compared to the horrifying consequences of a nuclear meltdown. The likelihood of a meltdown does not have to be a major consideration, it is the fact that it might occur at all that can be terrifying. Not only will it involve far more people than any single car accident, but it might also destroy the environment for years to come and carry with it unforeseen health damages that could affect the future of many people. Even decades after the Chernobyl accident, the area around Chernobyl is still not safe. Single car accidents might disrupt individual lives, which is bad enough, but a nuclear meltdown might change parts of our world for good. The magnitude of a hazard can be so severe that probabilities are less relevant. Furthermore, there is always the possibility that the experts might be wrong about the probabilities (Hansson, 2004), which might be an additional reason for people to prefer 'being safe than sorry.'

All these considerations show the reasons that give rise to the emotions that people have concerning certain risks, and they also show that we need imagination and sympathy to fully grasp the extent to which a risk might affect our lives and those of others. People who are convinced that the risks of nuclear energy are acceptable should give counter-arguments instead of dismissing the aforementioned concerns and emotions as irrational. It is not enough to show people that the quantitative risks of nuclear energy are lower than those of driving a car. People want to hear convincing arguments why nuclear energy is preferable to alternative sources of energy, as that is the relevant comparison in this case.[18]

The discussion of these examples is of course not meant to provide a stance, let alone exhaustive arguments against nuclear energy. Rather, it is meant to illustrate that risk emotions can be grounded in reasonable concerns and shed light on important moral values. As we saw in Chapter 3, purely quantitative approaches to risk might blur rather than clarify ethical issues. Moral emotions and intuitions help to grasp these ethical issues. For example, emotions such as sympathy, empathy and compassion can point out unfair distributions of risks and benefits, and indignation and resentment can point to moral digressions such as involuntary risk impositions.

5.6 Risk Emotions as a Solution to the Puzzle of Lay Rationality

Social scientists are struggling with how to deal with the fact that the risk perceptions of laypeople are largely based on emotions, as this seems to undermine the idea that laypeople might employ an alternative, legitimate rationality. This is what I called the 'Puzzle of Lay Rationality' at the end of the previous chapter, which I analyzed as follows.

Based on Chapter 4

P1: Laypeople include other concerns in their risk perceptions than experts.
P2: These other concerns are normatively justified.

—

Conclusion [prolay]: Laypeople's concerns are a normatively justified contribution to decision making about risk. (Conclusion [prolay] supports approaches to participatory technology assessment.)

Based on Chapter 4

P3: Emotions play an important role in laypeople's risk perceptions.
P4: Emotions are an unreliable source of decision making about risk.

—

Conclusion [antilay]: Laypeople's concerns are not a normatively justified contribution to decision making about risk. (Conclusion [antilay] denies Conclusion [prolay]; challenges approaches to participatory technology assessment.)

However, my alternative approach to risk emotions can provide a solution to this puzzle. Emotions are not necessarily a threat to rationality. While this idea follows on the DPT model, we have seen that emotion research provides us with an alternative framework, namely cognitive theories of emotions. This alternative understanding of emotions allows us to say that moral emotions are needed in order to grasp moral aspects of risk, such as justice, fairness and autonomy—aspects that cannot be captured by purely quantitative approaches such as cost-benefit analysis. By extension, we can then say that it is because emotions play a role in laypeople's risk perceptions that they have a broader approach to risk assessment that includes important moral values. Laypeople are more sensitive to the moral aspects of risk than experts who mainly rely on quantitative methods.

By departing from DPT and embracing a richer concept of emotions, we can reach different conclusions as to the role of risk emotions. We can replace P4:

> Emotions are an unreliable source of decision making about risk.

With P4'

> P4': Emotions are an important source of moral knowledge about risk.

Hence, rather than constituting a puzzle, emotions explain why laypeople have a broader, ethically more comprehensive understanding of risk than experts.

> P1: Laypeople include other concerns in their risk perceptions than experts.
> P2: These other concerns are normatively justified.
> P3: Emotions play an important role in laypeople's risk perceptions.
> P4': Emotions are an important source of moral knowledge about risk.
>
> —
>
> Conclusion [prolay enforced by emotion]: Laypeople's concerns and emotions are a normatively justified contribution to decision making about risk. (Lends additional weight to approaches to participatory technology assessment.)

Hence, the fact that the risk perceptions of laypeople involve emotions does not make them suspicious. To the contrary, we need moral emotions in order to have well-grounded insights into whether a technological risk is morally acceptable or not. For example, enthusiasm for a technology can point to benefits to our well-being, whereas fear and worry can indicate that a technology is a threat to our well-being; sympathy and empathy can give us insights into fair distributions of risks and benefits, whereas indignation can indicate violations of autonomy by technological risks that are imposed on us against our will. This approach can provide for a richer account of the importance of emotions in ethical reflection about risk than DPT. Understanding risk emotions as a source of moral insight is much more in line with Slovic's usual emphasis on the wisdom that is entailed in laypeople's judgments about risks. Rather than being biases that threaten decision making, emotions contribute to a correct understanding of the moral acceptability of risks.

5.7 Conclusion

In the literature on risk perception and acceptable risk, DPT seems to be an unquestioned paradigm. However, recent emotion-scholarship allows for a possible alternative interpretation of risk emotions. A different theoretical framework with a different view of emotions challenges the standard view of risk emotions as heuristics that are prone to be biased. Rather, emotions should be seen as an essential route to the proper understanding of the moral acceptability of risks. Based on such a different view of emotions, the empirical results of Slovic's studies on the 'affect heuristic' can be interpreted in a different light: emotions are a source of moral wisdom in thinking about risks. This interpretation is much more in line with Slovic's general work in which he stresses that public risk perception is based on legitimate concerns. This does not only mean a significant deviation from the mainstream view of emotions in the academic literature on risk perception and decision making under uncertainty, but it also has direct implications for risk policy from which emotions are generally banned or where they are seen as an unfortunate fact. My alternative approach to risk emotions implies that people's emotions should be explicitly invited into the arena of debates about acceptable risks, as an invaluable source of ethical wisdom and critical reflection. This will be the focus of part III of this book.

Notes

1. These are very crude characterizations of these positions as accounts vary, and there are also hybrid versions. That said, it is still common to characterize metaethical theories according to sentimentalism or rationalism. For example, a special issue of *Philosophical Explorations* that was devoted to the discussion of the meaning of neurological findings for moral philosophy was set up within the dichotomy 'sentimentalism' versus 'rationalism' (cf. Gerrans and Kennett, 2006).
2. As many anglophone philosophers, intuitionists use the notions 'moral' and 'ethical' interchangeably, which is what I also do in this book.
3. Exceptions are, e.g., Stratton-Lake (2002), Audi (2003), Huemer (2005) and Tropman (2009).
4. For example, see the symposium on *'Experiment and Intuitions in Ethics'* that was published in the journal *Ethics* in 2014.
5. The intuitionist Thomas Reid (1969b [1788]) discusses the role of affective states. According to him they result from and accompany ethical intuitions, and provide for motivation, but they do not play an epistemological role (cf. Roeser, 2009).
6. This can involve assessing general and particular moral judgments. This is similar to the Rawlsian approach of reflective equilibrium, which also does not offer a clear-cut methodology on how to balance principles and concrete judgments but requires an overall evaluation.
7. Cf. Chapter 3 for a short outline of Dancy's account.
8. I will discuss these two separate claims one by one in the following sections.
9. For extended objections against response-dependence accounts along similar lines, cf. Zangwill (2003) and Koons (2003).
10. For book-length defenses of non-reductive moral realism, cf. Shafer-Landau (2003), Cuneo (2007) and Enoch (2011).

11. I will discuss these ideas in more detail in Chapter 6.
12. The empirical findings and their interpretation are debated by psychologists and philosophers; cf. e.g., Kahane and Shackel (2010); Landy and Goodwin (2015) etc. However, my following argument applies even if the empirical findings are valid.
13. For example, by giving subjects (sociopaths, amygdala patients and controls) assignments to formulate moral judgments in accordance with the categorical imperative and testing their endorsement of these judgments. I leave the details of the tests to psychologists; for my argument, it suffices to realize that this is an empirical claim that can potentially be empirically tested.
14. I use value-judgment as a generic term here, including evaluative judgments about good and bad, normative judgments about right and wrong, and judgments about virtues and vices. I use the notions judgment and belief interchangeably.
15. There is a huge debate in the epistemology literature on how to analyze knowledge. Most approaches are framed around the notions justified true belief (JTB), but there is wide consensus that this is not sufficient as this can lead to counter examples, so-called 'Gettier-cases' (Gettier 1963 was the first to provide counter examples to an analysis of knowledge as JTB). To avoid entering into this complex discussion, I have included the formulation in component 4.
16. Note that Slovic does not so much speak about fear but about dread, which is an important factor in his work on the affect heuristic. According to Roberts, the notions 'fear' and 'dread' originally had a similar meaning but nowadays have different connotations (cf. Roberts, 2003, p. 199, n. 21). Roberts interprets dread as an emotion about an unavoidable situation. Unavoidability does not seem to be a necessary ingredient to dread in Slovic et al.'s studies. Hence, I will focus on Roberts' analysis of fear as I think that this comes closest to what Slovic means by 'dread.'
17. For a more detailed discussion of moral emotions concerning nuclear energy, cf. Roeser (2011b) and Taebi et al. (2012).
18. I will return to the discussion about nuclear energy in Chapter 7.

Part III

Emotional Deliberation About Risk

The epistemology of risk emotions developed in part II serves as the foundation for a novel approach to public deliberation about risk that is developed in part III. This new approach takes emotions as the starting point of debates about risk and as a source of critical reflection and deliberation.

6 Reflection on and with Risk Emotions

6.1. Introduction

In the previous chapter, I argued that emotions should play an important role in assessing ethical aspects of risks. This goes against common wisdom and the state-of-the-art in academic literature on risk and emotions, which see emotions as sources of biases of and disturbances to rational decision making. As argued in Chapter 4, not all supposed biases of emotions that have been identified in the literature really are instances of affect or emotion. Rather, I have argued that emotions are an indispensable source of insight when it comes to ethical aspects of risk.

However, this does not imply that emotions are an infallible guide. Emotions can help us to focus on certain salient aspects, but they can also lead us to overlook other aspects. This holds for all parties involved in assessing risks. For example, engineers might be misled by their emotions if their enthusiasm about a product leads them to overlook certain risks. Policy makers might be tempted to overlook the risks of a certain technology in favor of its potential economic prosperity to their region. If the public is ill-informed, it may only focus on the risks and overlook the benefits. All involved parties might be biased, and their emotions might reinforce those biases.

Emotions can cloud our risk judgments by blurring our understanding of *quantitative* information about risks. Furthermore, emotions can also bias our judgment of the *moral* aspects of risk. In Chapter 5, I mainly focused on moral emotions—that is, emotions involved in moral judgments about risks. However, the biases of emotions that were discussed in Chapter 4 mainly concern emotions that distort our access to scientific evidence on the descriptive aspects of risk and not the normative aspects. The main focus of the literature on risk and emotion is on emotions that constitute biases concerning descriptive aspects of risk. That is why most authors propose that if necessary, risk emotions could be corrected by rational and scientific methods. However, when it comes to biases in our moral understanding of risk, it is not obvious that pure rationality will help us out.

In this chapter, I will explore the methods that can be effective in critically reflecting on emotional responses to risk. I will argue that there is need for a division of labor: if emotions bias our quantitative understanding of risk, we indeed need proper—and accessibly presented—*quantitative* information, as has been argued by Slovic, Sunstein, Kahnemann and others. However, regarding emotion-based biases concerning the *moral* evaluation of risks, we need a different approach, targeted at critical moral thinking. I will argue that here emotions can also play an important role.

As I argued in Chapter 5, the standard approaches in ethics see emotions as follows: rationalists argue that we should correct our emotions using reason, while sentimentalists argue that emotions should rule. I will take a different line. In this chapter, I will argue that the cognitive theory of emotions that I developed in the previous chapter allows for the idea that emotions themselves have critical potential. Reason and emotion should be used to critically reflect on each other, but emotions should also be used to critically examine other emotions, by trying to understand different perspectives through sympathy and empathy. Emotions can themselves be a source of critical reflection and deliberation. For example, those who benefit from a technology should try to understand the perspectives of those who are potential victims of the technology. Other-regarding, altruistic emotions can help us criticize our initial egoistic emotions, for example in the cases of a NIMBY response or our attitude to climate change. I will specifically focus on sympathy and compassion to assist with critical reflection on and the proper formation of reliable risk emotions.

6.2 Emotional Biases Concerning Quantitative Aspects of Risk

As I have argued in Chapter 4, emotions are overly crudely characterized by Dual Process Theory (DPT), which sees emotions as opposites of reason and rationality. In Chapter 4, I argued that not all supposed biases in thinking about risk and uncertainty are genuine biases, and not all biases are due to emotions. Further, in Chapter 5 I argued that emotions can highlight important moral aspects of risks. One caveat is that, as with our other capacities of perception, judgment and decision-making, emotions can also be misguided. This may be especially pertinent in the case of risk-emotions, because the information is complex and inherently uncertain, and because a lot can be at stake given the impact technologies can have on people's lives. The question is how we can examine emotions involved in biases concerning risks and if necessary, correct them.

Emotional responses to risk can be directed at 'factual' (descriptive, empirical, scientific, quantitative) aspects as well as at moral aspects of risks. In both cases, they can be justified or biased (and this can of course be a matter of degrees). For example, factual aspects of risk can give rise to fear, and this can be a rational or an irrational response to factual aspects of risk. One example of a rational response is the fear of snakes, which can lead us to avoid snakes. An example of an irrational response is the fear of flying, which is not easily justified given the safety statistics of air travel. People's fear of flying might lead them to wrongly perceive the purely quantitative level of risk. Cass Sunstein calls the phenomenon where people's fear of a negative outcome of a technological activity blinds them to its low probability, 'probability neglect'. In these cases, scientific evidence is needed to correct emotions and to put things in perspective.

This is complicated by the fact that some risk-emotions function like stereotypes or phobias and can be hard to penetrate with factual information. For example, fear of flying may still persist, even in the light of evidence about the safety of air travel. Factual information needs to be presented in an emotionally accessible way in order to be able to correct misguided risk-emotions that are directed at factual aspects of risk (cf. Buck and Ferrer, 2012). One way to provide balanced and emotionally accessible information might be to point out the benefits of a technology in cases where people primarily focus on small risks.

Sandman proposes the following strategies for correcting risk-emotions: 1. teach people about hazards; 2. make serious hazards outrageous; and 3.

> we have to stop contributing to the outrage of insignificant hazards. As long as government and industry manage low-hazard risks in genuinely outrageous ways—without consulting the community, for example—citizens will continue to overestimate these risks and activists will continue to mobilize against them.
>
> (Sandman, 1989, p. 49)

Sandman seems to suggest that outrage can actually be created or enhanced by concealing information, and it can be taken away by involving the public. Involving the public creates trust (cf. Slovic, 1999; Asveld, 2009). Experts often tend to not inform the public about scientific information on risks because they think that the public will not understand the information, or because they are afraid of lawsuits should their estimates turn out to be wrong, or should a hazard arise which they claimed was highly unlikely. However, if policy makers and experts downplay risks and uncertainty and an unlikely event nevertheless occurs, then this can actually undermine trust (cf. van Asselt and Vos, 2006). A lack of trust in authorities and technology experts can lead people to opt for a precautionary 'better safe than sorry' approach toward technology. Hence, it is also in the interest of experts to be transparent and involve the public. If experts are convinced that a certain technology is worth undertaking, they should share their knowledge about the quantitative risks and benefits as well as their ethical concerns with the public. Furthermore, various authors have emphasized that trust also involves emotions (Möllering, 2001; Zinn, 2008; Alfano et al., forthcoming). This is another reason for policy makers, experts and members of the public not to eschew emotions, as they can point to important ethical considerations as well as being an important ingredient in a relation of trust between different stakeholders.

Furthermore, it is not necessarily the case that whenever probabilities are low, emotional resistance such as fear is irrational. Some risks have such catastrophic effects that probabilities become less significant, especially in the face of available alternatives. This can influence people's level of fear toward a technology. Taking nuclear energy as an example, a nuclear meltdown might change large parts of our world for good, even though it is extremely unlikely to happen. There are alternative sustainable energy sources available that do not have such catastrophic side effects (see the examples discussed in the previous chapter). This points toward the importance of ethical aspects of risk in the assessment of which emotions are indispensable, as I argued in Chapter 5. Nevertheless, emotions concerning ethical aspects of risk can also be biased.

6.3 Emotional Biases Concerning Ethical Aspects of Risk

In cases where emotions blind us to empirical facts, they should be corrected by scientific methods. However, as discussed throughout this book, the notion of risk is not only a quantitative but also an evaluative notion. In the previous chapter, I argued that emotions are necessary in order to obtain moral knowledge concerning risks. Yet, this does not imply that emotions are infallible as a normative guide. While emotions can help us to focus on some morally salient aspects of risks, they can also cause us to overlook other aspects. In the case of risk-emotions directed at moral aspects of risks, these may or may not be justified.

Emotions can point out important moral considerations, but they can also be notoriously misleading. Prime candidates are fear and disgust. In the previous chapter, my alternative framework of emotions established why moral emotions such as sympathy, empathy and indignation can provide important insights into the moral acceptability of risk. But fear and disgust are more complicated. Fear and disgust might just reflect our unfounded prejudices and phobias. Fear and disgust are less clearly focused on moral aspects of risk; they can also be responses to perceived threats that might be based on wrong factual information. Even in the light of contrary moral or factual evidence, we might still feel fear or disgust (cf. Sunstein, 2005, for the irrationality of fear).

Jonathan Haidt and Jesse Graham (2007) claim that disgust is a predominantly conservative emotion. They give the example of the disgust that conservatives feel toward homosexuals. Leon Kass (1997) argues that disgust concerning developments in biotechnology such as cloning is a justified response that favors traditional forms of procreation and family ties. However, Dan Kahan (2000) has argued that disgust can also figure in a progressive moral framework. For example, one can be disgusted at racists, sexists and homophobes.

I would like to argue that in the context of risky technologies, disgust can be biased toward a conservative status quo, but it can also point toward the morally ambiguous status of artifacts of synthetic biology, for example. Ethical objections to new technologies such as cloning, human-animal hybrids, cyborgs or brain implants are often linked to reactions of disgust and 'uncanny' feelings. These feelings can point to the unclear moral status of these creatures, to our unclear responsibilities for them and to the worry that they might develop in an unforeseen way. These are ethical concerns that need to be addressed when developing and dealing with these types of new technologies; and disgust can point us toward morally salient issues (cf. Miller, 1997; Kahan, 2000 on the rationality of disgust).

Similarly, fear of risky technologies can be notoriously misleading, but it can open our eyes to dangers that we may not otherwise be sensitive to. Fear can also make us aware of the morally problematic uncertainty that risks introduce, such as potentially large-scale catastrophic risks. Experts might feel responsible for and worried about the technologies they develop. Fear might point to concerns about the technologies' unforeseen negative consequences. Fear and disgust can be warning signs, making us aware of the moral values involved in new technologies. When fear and disgust can sustain reflection, they should inform our judgments. Further on in this chapter, I will argue that this reflection can itself involve emotions. However, I will first discuss the complexity of fear in the context of uncertainty in the following section.

6.4 Problematic Risk Emotions: Fear in the Light of Uncertainty

In the literature on risk emotions, the emotion that gets most widely discussed and criticized for its potential to disturb rational thinking is fear (also referred to as dread or anxiety). One example is Cass Sunstein's book *Laws of Fear* (Sunstein, 2005), which is devoted to the discussion of the detrimental effects that fear can have on decision making about risk. In the previous chapter, I argued that fear can be a reasonable response to risk. In this section, I will zoom in on fear and anxiety in the context of uncertainty, as fear may be a less reasonable, albeit understandable, response to circumstances in this context. I will provide a taxonomy of different risk and uncertainty domains, as well as an analysis of how these domains have different impacts on experiences of uncertainty. Furthermore, I will discuss different aspects of uncertainty-related anxiety, such as the aforementioned ethical aspects of risk and how they may figure in perceived uncertainty and concomitant anxiety. I will also provide a short explanation on why we may not be particularly well prepared to deal with uncertainty, which can result in fear and anxiety.

People can be risk seeking and risk averse, but they can also be uncertainty averse (e.g., Epstein, 1999; also cf. literature on the Ellsberg paradox and ambiguity aversion, cf. Ellsberg, 1961). In other words, if risk is at least the possibility of an unwanted effect, then it is not only the unwanted effect that people may want to avoid but also the uncertainty inherent to the possibility. This can lead to fear and anxiety. I will discuss three major areas in which fear related to uncertainty can occur.

Type 1. health risks;
Type 2. voluntarily invoked life-goal risks; and,
Type 3. technological risks.

People can fear health risks; for example, they may fear the results of a health screening (type 1). People can also fear the uncertainty of voluntarily invoked risks related to one's life goals, for example concerning a job application or a research grant application (type 2). In terms of type 3, technological risks, an example would be the risk of a nuclear meltdown. One such example is when people feared imminent danger in the aftermath of the Fukushima tsunami. But there is also the more general fear of possible hazards emerging from a technology.

What the three types have in common is that there is a lot at stake, even though the timing of when the outcomes manifest themselves may be uncertain. There may be uncertainty both about the outcomes as well as the timing of the outcomes. This could make the uncertainty even harder to bear.

Let us now look at what distinguishes these types from each other, as this could provide us with a taxonomy and further understanding of uncertainty and concomitant fear. To that end, I will look at several salient qualitative dimensions of risk commonly discussed in the literature on ethical and social aspects of risk (also see Chapters 3 and 5). These qualitative dimensions of risk are:

- whether risks are collective or individual;
- how risks and benefits should be assessed, measured, compared and distributed;
- the status quo;
- voluntariness; and
- natural versus human-made risks.

As discussed in Chapter 3, these are qualitative or *ethical dimensions* of risk, which are not part of formal approaches to risk in rational decision theory, but they do play a major role in people's risk perceptions, and they can be ethically justified. As discussed in previous chapters, cost benefit analysis, the dominant approach in formal methodologies for risk assessment, focuses on *collective* risks and benefits and does not take the impact of risks and benefits on individuals into account. By focusing on aggregate levels of risks and benefits, it also overlooks issues of the *equity and fair distribution* of these risks and benefits within a given population. It is often far from clear how and on what scale risks and benefits should be measured and *compared* (cf. Espinoza, 2009 on the problem of incommensurability in the context of risk). Attachment to the *status quo* can be a bias if the status quo is bad and can be easily improved at no risk. But if there are good reasons to value the status quo, there might be good reasons to be cautious about risks of a new development, even though it may provide for improvements. *Voluntariness* relates to the morally important concept of autonomy, but it does not figure in the dominant consequentialist, technocratic approaches to risk such as cost benefit analysis. *Naturalness* tends to be an important factor in laypeople's risk perception, but it is not distinguished as such in formal risk approaches. Naturalness might be a more contentious issue. On the one hand, it seems more consistent with a sustainable, health-conscious lifestyle. On the other hand, something that is natural is not automatically less risky than something that is artificial or technology based (cf. Hansson, 2003). There are many dangerous substances to be found in nature, such as uranium or asbestos, and most substances can be dangerous in the wrong dose. Furthermore, technologies can protect us from natural hazards, for example dikes and walls.

In Chapter 5, I discussed how emotional responses to these qualitative and ethical dimensions can be justified. In the following discussion, I will analyze the way in which these dimensions play a role in the three different types of risk and their impacts on anxiety and experienced uncertainty, and whether the ethical dimensions help to understand or even justify these responses.

Collective vs. Individual Risks

With the exception of pandemics, health risks are predominantly experienced individually. Similarly, risks related to life goals are typically individual as well. In contrast, technological risks such as those related to energy production, infrastructure, climate change or the use of new substances, such as nanoparticles, are often collective as they are introduced to society on a large scale. In cases of collective risks, the options for individual action are less clear than in cases of voluntarily invoked, individual life goal risks. This is related to the dimension of voluntariness that I will discuss below.

Risks Versus Benefits

In risk types 1 and 2 (health risks and voluntarily invoked life-goal risks respectively) presented above, the benefits of a potentially risky situation are clear. People aspire to be healthy, although many people have to deal with the unfortunate reality of illness. Getting a new job or a major research grant is a very desirable situation and is thus worth going through the risk of not getting it. However, in type 3 (technological risks), the benefits are often less clear to society, which explains why the risks are seen as more controversial. Risk types 1 (health) and 3 (technology) have clear negative impacts. Risk type 2 (life goals) could mean that one has to find a different career, but it could also be less dramatic—for example, where a new job or grant could improve one's current work situation although the situation is acceptable as it is.

Status Quo

Another important question when thinking about the acceptability of risk is what the *status quo* is. In type 1, the *status quo* is the best-case scenario—that people are healthy. Medical treatments primarily serve to restore the *status quo*; and medical tests assess whether someone is healthy or in need of medical treatment. In the case of type 2, people have nothing to lose when they apply for a grant or a new job other than the time investment and the stress that the uncertainty can bring. In type 3, the *status quo* is not to introduce new technologies. Technologies are developed to improve well-being, which is usually an improvement of the *status quo*. However, the perception of controversial technologies is often that they entail seemingly unnecessary risk. The widespread phenomenon of *status quo* bias (cf. Samuelsen and Zeckhauser, 1988) may explain the aversion people have to technologies: the *status quo* is seen as the normal or natural state. Sociological studies of technology acceptance show that what are nowadays perceived as non-controversial technologies were at the time often perceived as threatening and dangerous. Examples include trains and telephones. Similar patterns of reluctance occur again and again when a new technology is introduced (cf. Schivelbusch, 1986 [1977]). However, it can be a misconception that the status quo preceding the introduction of a new technology is risk free (cf. Sunstein, 2005). On the other hand, some technologies turn out to be very risky and, once introduced to society, develop a momentum of their own. Examples include nuclear weapons, which played a crucial role in the arms race during the Cold War, and nuclear energy, with its concomitant unsolved problem of nuclear waste (cf. Taebi et al., 2012).

In summary, in type 1, the *status quo* (health) is the best-case scenario; in type 3, the *status quo* tends to be perceived as the best-case scenario, even though this might be unjustified; while in type 2, the *status quo* is presumably the worst-case scenario (except for the time investment and stress during the application).

Controllability, Voluntariness and Natural Risks versus Human-Induced Risks

I will discuss these notions together as they are closely related in the context of my discussion. The common factor among the three types of risk (health risks, life goal risks and technological risks) is that they are characterized by an experienced and factual lack of control. However, one difference between them is the even more involuntary nature of health risks versus the human-induced, and in that sense, avoidable, risks of technology. As health risks are in part natural and unavoidable, they require a degree of coping skills and fatalism. However, in the case of technological risks, people may wonder why they put themselves in risky situations at all. Fatalism might not be an appropriate response if risks are avoidable. That said, if technological risks result from collective actions, individuals may still feel helpless and disenfranchised. Cases of problems brought on by collective action point to the complex interrelationships between individual and collective stakes, benefits and responsibilities. Hence, in type 3 cases too, people might experience anxiety arising from uncertainty in a similar way as in type 1. Similarly, in cases of imminent threat, issues of causal responsibility might be less relevant as they do not make a difference at that stage. This becomes clear in the case of type 2, where we experience anxiety brought on by uncertainty, even if we ourselves have partially initiated that uncertainty. This is especially true if our fate is out of our hands and depends on the decisions of others.

Let us further complicate matters by acknowledging that distinguishing the cases according to degrees of voluntariness is problematic. For example, an unhealthy lifestyle can cause health problems. Lifestyle-induced health risks can then be seen as a mix of types 1, 2, and 3: they concern health (type 1); they are self-invoked and human-induced (type 2); and they may involve technology, for example driving a car instead of riding a bicycle (type 3). However, the link between lifestyle and health risks is more tenuous in some cases than it is in type 2 cases. This is because health problems may have multiple causes and it can be hard to relate a health problem to lifestyle in individual cases. Further, even type 2 cases are often only partially initiated by us as, arguably, circumstances may push us to engage in uncertain activities, such as applying for a new job. This is paralleled in the case of the link between lifestyle and health risks where external circumstances such as socio-economic background, education level and health promotion in the society one lives in may be influential factors. There are factors though that may be, to various degrees, within one's own control, such as the availability, accessibility and affordability of healthy food, awareness of risks associated with smoking and opportunities for physical exercise. This illustrates that individual choices and systemic circumstances are intertwined and disentangling them is a bit artificial. That said, I do nonetheless disentangle them, as the different types are to some extent more or less voluntary, albeit not purely one way or the other.

The table below summarizes the taxonomy of risk types according to their ethical and qualitative dimensions as discussed above.

Table 6.1 A taxonomy of uncertainty types and their (perceived) qualitative dimensions of risk.

	Type 1 (health)	*Type 2 (life goals)*	*Type 3 (technology)*
Collective	–	–	+
Value of benefits	+	+	–
Status quo attachment	+	–	0
Voluntariness	–	+	0
Human-induced risk	–	+	+
Controllability	–	–	–

In conclusion, uncertainty-related anxiety can occur in all three types of risk, and one feature that they all share is a lack of control. However, there are some differences between them. In type 1, the situation is out of one's hands from the start, whereas in type 2, when trying to achieve an aspired life goal, the situation was initially in one's own hands, but is subsequently out of one's hands while awaiting an external decision. In the case of type 3, lack of control, uncertainty and related anxiety can lead to resistance to the technology.

The question remains as to whether uncertainty-related anxiety is an irrational, biased state that needs to be corrected, or whether it is an understandable phenomenon that might even reveal insights that we might not be aware of without it. I believe that uncertainty-related anxiety is ambiguous in that respect, as I will now argue.

From an evolutionary perspective, we are not prepared for any of the three types of risks that I have discussed.

Type 1. health risks: in the past, maybe even up to 100 years ago, we had no sound knowledge about health risks nor the sophisticated medical diagnoses that we have today. If they did exist at all, it was not at any significant level.

Type 2. voluntarily invoked life-goal risks: humans have exposed themselves to these risks throughout history, but in the past, it was the privilege of a small class of people. In contemporary, individualistic Western society, designing one's life and destiny is pervasive in all classes, albeit in different forms and to different degrees.

Type 3. technological risks: we have experienced technology-associated risks since the industrial revolution, but genuine techno-science and the wide spread of technologies in most areas of our lives, with their concomitant risks and uncertainties, have mainly arisen over the last 100 years.

Hence, an evolutionary perspective can indicate that it is not surprising that we have problems dealing with risk and uncertainty, which can lead to stress and anxiety. In the case of uncertainty-related anxiety, the most a person might be able to achieve is to be able to cope with the uncertainty (cf. Little and Halpern, 2009, on coping). What makes it difficult to cope with uncertainty is that the options available to the agent for action are unclear as the outcomes can go either way, and it is hard to predict what will happen. What one presumably wishes for in such a situation is to regain control. *Withdrawing* from the situation can then be experienced as a solution. But withdrawal is neither a real option nor an attractive option in any of these three cases.

For example, in the case of a type 2 risk, if someone withdrew a job or grant application, she would never know whether she might have gotten the job or grant or not. However, she might learn from her anxiety level that applying for this job or grant comes at too high a price, namely the uncertainty involved is too hard to bear. In that case, she might withdraw her application, thereby regaining control over the situation. However, in some cases, people suffer so severely from uncertainty-related anxiety that they cannot stand the pressure anymore. Indeed, there are cases where people commit suicide while waiting for the result of a competition. In France, Michelin star chef Bernard Loiseau committed suicide shortly before he was to hear if he would keep his three Michelin stars. The pressure of the negative reviews and rumors that he might lose one of his Michelin stars became unbearable for him. In the end, his restaurant kept the star.[1]

In the case of the result of a health examination, withdrawal is also not a clear option. A possible response would be for someone to withdraw from medical diagnosis altogether, maybe turning to alternative medicine.

In the case of technological risks, there are people who try to withdraw and lead an alternative lifestyle. However, one of the intricate ethical aspects of many technological risks is that they can also affect people who choose not to make use of the risky technology. Technological risks can have far-reaching effects, for example on the environment or through accidents that can also affect people who are not directly involved in using the technology.

Hence, in some cases withdrawing from the source of the uncertainty in order to regain control might be an option; in other cases, coping might be the best available option. In the case of technological risks, other alternatives could be joining an environmental movement or trying to change policy or technology development in other ways. As discussed in Chapter 2, in the last few decades, many approaches have been developed in participatory technology assessment that try to include laypeople in constructive decision making about risky technologies and in technology development itself. These approaches can take people's fears and concerns as input in order to come to better decisions about responsible innovations, an idea I will pursue further in Chapters 7 and 8.

To summarize, uncertainty-related anxiety can point to important ethical aspects of risk, but it can also be problematic in that it may lead to overreaction, stress and difficulty in putting things in perspective. This can partially be explained by a lack of general and evolutionary preparedness for the kinds of uncertainty we face in contemporary society. However, uncertainty-related anxiety can also be related to ethical aspects of risk that deserve further examination and critical reflection.

6.5 Correcting Emotion Through Emotion

In the preceding sections, I discussed emotional responses to risk that can be biased in terms of both scientific as well as ethical aspects. Looking at fear and disgust, these are especially prone to be based on stereotypes, phobias and biases. However, I also argued that fear and disgust can point out important ethical aspects of risk. The question is how to distinguish between emotions that are helpful versus those that are detrimental to our ethical judgments, and how to critically reflect on them in decision making about risks. In what follows, I will discuss how we can correct misleading risk emotions directed at evaluative aspects of risks.

Rationalists would propose that we need rational methods to critically assess emotions. However, as I have argued in the previous chapter, rationality or reason is not necessarily superior to emotion, especially in the context of moral judgments. I propose that emotional moral judgments can be justified if they can sustain reflection, but reflection is itself a process in which emotions can play an important role. We can use the reflective and critical potential of emotions, their capacity for shifting points of view and caring for the well-being of others to critically examine other emotions.

Emotions can be reflective. For example, we can use them to assess our confidence in our other emotions and moral judgments. If we feel insecure about our moral viewpoint, we may doubt whether we are right. If we feel outrage at the violation of a moral norm, we are confident of that norm. But in the light of fundamental disagreement, we can make efforts to draw on our empathy and sympathy, put ourselves in the other person's shoes and feel compassion for that person. We might even consider reassessing our emotional moral belief by looking at the situation from different points of view. Furthermore, some moral emotions might be more susceptible to doubt than others. Moral emotions in dilemmatic or complex situations are more fallible, which can be expressed by feelings of desperation about whether we made a right judgment, or by feelings of being torn between two different emotions. Emotional responses can be informative in assessing whether an emotion corrects or corrupts our initial moral judgment. We might feel uncomfortable and that we are cheating when we let go of an initial judgment based on an emotion, but we might also feel forced to reconsider our initial judgment and feel relieved if we align our judgment and our feeling about a certain issue. While the former feeling might point to a corruptive emotion, the latter might point to a corrective emotion.

Michael Lacewing (2005) makes a similar argument based on ideas from psychoanalysis. He argues that we need to examine our emotions through 'emotional self-awareness', which involves three factors:

1. feeling the emotion;
2. being aware of feeling it; and
3. normally, feeling a second-order emotional response to it.

He adds a 'dispositional fourth', an openness to emotions, which he explains as 'a readiness to feel and acknowledge what emotions one has' (Lacewing, 2005, p. 68).

Through this process of emotional self-awareness, we are able "to detect our anxiety which raises the possibility that our emotional response to the situation is being driven by defense mechanisms" (Lacewing, 2005, p. 73). This is important because "[e]motions that are the product of defense mechanisms are not appropriate evaluative responses to the world" (Lacewing, 2005, p. 73). A purely rationalist approach runs the danger of a form of intellectualization that "defends against anxiety partly by working with denial, isolation, or repression to simply not *feel* the emotion that arouses anxiety, and partly by using various means of avoiding the emotion's implications and personal significance" (Lacewing, 2005, p. 75). As Lacewing emphasizes: "[n]ot *feeling any emotion does not mean one's thinking is undistorted*" (Lacewing, 2005, p. 76; italics in original). In other words, suppressions or rationalizations of emotions can also be distortions. Lacewing argues that even in cases where emotions are disruptive, it can be important to examine why one feels that emotion instead of just laying it aside. In these cases, the emotional self-awareness can be "detached" but still "engaged" (Lacewing, 2005, p. 80).

Let us apply these ideas to emotions in the context of risk. When thinking about whether we find a risk morally acceptable or not, we should reflect on our emotions about the risk as well as on our emotional responses to these emotions. Could our fear of a given technology be sustained by further reflection? Does this fear seem genuine? By using emotions such as sympathy and empathy, we can take a more general perspective and try to feel from the perspective of people who may be victims or beneficiaries of that technology. A question to ask may be whether we think that this technology is acceptable overall to society or not. This kind of emotional reflection may reveal that we feel upset about a certain technology because we do not wish to undergo its negative side-effects ourselves, even if we think that overall it is a desirable technology for society. This might indicate that we are driven by egoistic views rather than by genuine moral concerns about that technology. This would be an example of the NIMBY problem: I am not against the technology *per se*, I just do not want it 'in my backyard'. But of course, in the case of a technology that is overall desirable but has certain negative side-effects, we know that these side-effects will affect some people, and it is only fair that everybody be affected by them at one point or another. If it is a genuine case of egoism, then higher-order emotional reflection can point this out and help us overcome our egoism.

Alternatively, one's unease with an overall desirable technology might point to better ways to deal with negative side effects than is initially proposed. In this case, the feeling of unease should be taken seriously since it can point to morally important considerations such as the unfair distribution of risks and benefits, or to the fact that the risks are imposed on some people without giving them the chance to have a say in what is happening.[2] It can also point to other, less risky and comparatively equally beneficial alternatives; or to certain side effects that, though unlikely, are so catastrophic that they are simply unacceptable given their potential victims. A useful test to check our emotional motive would be to consider our emotional response if we abstract from the idea that we ourselves are the potential victims and imagine that another person is the victim. If we still think that it is unfair, it is apparently not just an egoistic emotion.

This is of course tricky because one of the ways in which we can invoke emotions such as sympathy is to understand the moral value of a situation concerning another person by imagining oneself in the shoes of that person. Doing this makes it easier to see what might be wrong in that situation. And now I am proposing doing the opposite, which may be asking too much of our imaginative capacities. After all, we tend to care more about the well-being of our nearest and dearest than of distant others. On the other hand, this concern might be based on a too narrow view of moral emotions such as sympathy. Sympathy can also be directly felt for another person, without the need for a detour through our personal perspective.

The claim that emotions are necessary for moral judgments about risk does not entail that they are infallible. As with all sources of knowledge, emotions can misguide us. But whereas we can use glasses or contact lenses to correct imperfect vision, there are no similar tools to correct our emotions. However, emotions themselves can have critical potential. Sympathy, empathy and compassion allow us to take on other points of view and critically reflect on our initial emotional responses. Hence, we can correct misguided moral emotions via emotions.

6.6 Correcting Reason through Emotion

Furthermore, apart from emotions being able to correct emotions (section 6.5) and reason being able to correct emotions (section 6.2), emotions can also correct reason. The idea that emotions can be corrected by reason is commonplace in academic literature as well as in public debates. The idea that emotions can also correct emotions has not previously been studied in the context of risk, and only a few philosophers have discussed it in the broader context of ethical reflection. The idea that emotions can correct reason is hardly discussed at all even though some authors who write critically about risk-emotions still emphasize that without emotions, we would be without any sound guidance. They often invoke Damasio's work on the so-called 'somatic marker-hypothesis', as in this passage from Loewenstein et al.:

> Emotional reactions guide responses not only at their first occurrence, but also through conditioning and memory at later points in time, serving as somatic markers. Patient populations who lack these markers not only have difficulty making risky decisions, but they also choose in ways that turn their personal and professional lives to shambles. Thus, feelings may be more than just an important input into decision making under uncertainty; they may be necessary and, to a large degree, mediate

> the connection between cognitive evaluations of risk and risk-related behavior.
>
> (Loewenstein et al., 2001, p. 274)

As discussed in Chapters 2 and 3, quantitative approaches to risk such as cost-benefit analysis have been under severe attack (e.g., Fischhoff et al., 1981; Shrader-Frechette, 1991; Slovic, 2000; the contributions to Asveld and Roeser, 2009). As I have argued in the previous chapters, quantitative approaches need to be corrected and complemented by the ethical insights present in laypeople's risk-related intuitions and emotions. Hence, while Sunstein warns us against 'probability neglect' (cf. section 6.2), a technocratic approach may lead to what I would like to call 'complexity neglect'. By merely focusing on easily quantifiable considerations, such as annual fatalities, for example, we may overlook other morally relevant considerations which could be revealed through emotions such as fear and sympathy. As argued in Chapter 5, risk-emotions can be based on reasonable concerns. These concerns should be taken seriously in debates about the acceptability of technological risks. In other words, one should aim to avoid both probability neglect and complexity neglect, which means that quantitative and moral approaches should be used in combination. Quantitative information is necessary, but it is not sufficient for a moral evaluation of risks. For the latter, we need ethical approaches, and as I argue in this book, this also means paying attention to emotions.

Nevertheless, even if one accepts that emotion can be a source of ethical insight, a common perception is that emotions are notorious for being more misleading than other mental abilities. Disruptive emotions such as mass panic, rage, jealousy or uncritical enthusiasm lead people to do things that are morally wrong. However, as significant as these examples are, there are also examples of emotions that contribute to our moral insight and complement and even correct rationality. To reiterate, beings that are purely rational and have no emotions are unable to make proper practical judgments, especially when it comes to concrete moral judgments in particular situations (Damasio, 1994). Furthermore, it is wrong to think that only purely rational moral beliefs can be justified and critically reflected upon—purely rational beliefs can be misleading as well, and emotions can correct them. Nussbaum even thinks that "emotions are not only not more unreliable than intellectual calculations, but frequently are more reliable, and less deceptively seductive" (Nussbaum, 1992, p. 40).

This relates to the economist Robert Frank's ideas, who argues that altruistic emotions can solve rational choice problems such as free-riding, that is cases whereby someone does not cooperate, but takes advantage of others' cooperation (Frank, 1988). Sympathy and fellow-feeling can help overcome 'cold-blooded' supposedly rational egoism as well as selfish feelings, and promote cooperation instead. A case in point is people's attitudes toward climate change whereby someone's concern for the environment can trump rationally calculated behavior that contributes to climate change. Using a purely rational calculation, one could justify one's own free-riding, polluting behavior as an individual contribution that has little impact. We know from rational choice theory and game theory that if everybody thinks this way, this is what creates the problem of exhaustion of resources. Still, an individual could conclude that his decision to not travel by airplane will not, in itself, change other people's decisions regarding flying. He therefore rationalizes his own environmentally bad behavior.[3] However, someone who substantively cares about the environment could conclude that environmental concerns are more important than arguments for free-riding, even if the latter are compelling rationally. In this case, emotions are recalcitrant: they go against a rational judgment and persist. The common idea is that these recalcitrant emotions should be corrected by rationality. But the example of climate change demonstrates that recalcitrant emotions can make us aware of important moral insights. Emotions can enable us to make better moral judgments by helping us to reverse our selfish, rational judgments. In the case of climate change, altruistic emotions can help us care about the well-being of people who are distant in space and time, and these emotions can lead us to make personal sacrifices by adopting a more sustainable lifestyle.[4]

Emotions are not infallible, but they *can* lead us to see what is morally right, and they are often better in doing so than our purely rational judgments. This is supported by empirical studies by Paul Slovic concerning donations for starving children in Africa. People donate higher amounts when their compassion is invoked by the picture of one starving child. They donate lower amounts when there are two children, and they donate the least when they read statistical information about millions of needy children in Africa. In Slovic's words:

> People don't ignore mass killings because they lack compassion. Psychological research suggests it's grim statistics themselves that paralyze us into inaction.
>
> (Slovic, 2007)

The title of Slovic's article is 'numbed by numbers'. Rational, statistical information can make us numb; it indicates that the situation is hopeless. In contrast, the destiny of a single child can successfully evoke strong feelings of compassion and charitable behavior. Slovic et al. (2004, pp. 320, 321) argue that affect can be more suited to convey meaning than sheer numbers. They give examples of works of literature and art that let us better understand the horrors of the Holocaust and other catastrophes than statistics. The philosopher Martha Nussbaum (2001) also emphasizes that sympathy can broaden our 'circle of concern', for example through reading works of fiction. Emotions can be a source of critical reflection and deliberation, for example by appealing to narratives and the imagination, in which art can play an important role. I will come back to the possible role of art for emotional-moral reflection on technological risks in chapter 8.

6.7 Conclusion

Emotions are necessary for moral knowledge, but they are no guarantee for success. Emotions are not infallible guides to knowledge, but this is the case with all our potential sources of knowledge. Even a rationalist cannot claim that reason always gets it right. In this respect, all sources of knowledge are in the same boat. However, emotions are often perceived as more notoriously misleading than other mental abilities. I think that this is a mistaken view. To the contrary, purely rational beings without emotions are unable to make proper moral judgments, especially concrete moral judgments in particular situations, as is shown by the famous studies by Antonio Damasio (1994). Risk-emotions may have blind spots, but without emotions we would miss out on important evaluative aspects of risk. Even a suspicious emotion such as disgust might point to the ambiguous moral status of, for example, clones and human-animal hybrids.

When thinking about risks, emotions and science should be in balance: where science can inform us about magnitudes, emotions can inform us about moral saliences. Both kinds of information are crucial if we want to make well-grounded judgments about acceptable risks. Scientific methods with which to measure risks are important corrections to emotions if people tend to ignore scientific evidence because they are consumed by their emotions. On the other hand, emotions provide us with awareness of ethical aspects of risk that are not included in quantitative approaches to risk. Furthermore, emotions can play a role in the critical examination of our moral views, including in the context of risk. Hence, as is widely acknowledged, emotions can be corrected by reason, but emotions can also correct emotions, and emotions can also correct misguided rational beliefs. These ideas will be developed further in the following chapters, where I will examine how emotions can play a constructive role in public deliberation about risky technologies.

Notes

1. Cf. http://en.wikipedia.org/wiki/Bernard_Loiseau for information. Thanks to Peter Kroes for drawing my attention to this case as an example of uncertainty-related anxiety.
2. Basta (2012) argues that in the case of debates about siting of risky technologies, supposed NIMBY responses are actually frequently due to procedural and/or distributive unfairness.
3. Cf. Lorenzoni and Pidgeon (2006, p. 85), where they describe this attitude among people whom Pidgeon and colleagues have interviewed in another study; also cf. Moser (2010, pp. 34–35) for similar findings.
4. I will come back to the example of climate change in Chapter 8.

7 Participation with Emotion

7.1 Introduction

In Chapter 2, I sketched the current dominant approaches to decision making about risk. Technocratic approaches to risk ignore emotions and values. Populist approaches take emotions as endpoints of debates. Participatory approaches involve the public in deliberation about risks. However, they do not explicitly address emotions, and thus they may not do full justice to the moral concerns of the public. As argued in Chapter 2, this is also reflected in practical debates that draw on these approaches. The fact that these approaches do not pay sufficient attention to emotions and moral values may explain why many public debates about risk result in stalemate, and they leave out important resources for democratic decision making.

The new approach to risk emotions developed in part II provides the theoretical foundations for a richer, more fruitful participatory approach. It does this by letting emotions play an important role in decision making about risk as they can point to important moral values. In the previous chapter (6), I discussed how emotions can be the object as well as the subject of critical reflection. In this chapter, I will discuss how risk emotions can be integrated in political participation and decision making about risk. Rather than ignoring emotions as in the technocratic approach, or taking emotions as endpoints of discussion as in the populist approach, or including the public only to the extent that it provides 'rational arguments' as in conventional participatory approaches, the approach developed in this chapter states that emotions should be the starting point of risk debates. By asking people what triggers their emotions, substantive moral considerations underlying the emotions can be made explicit. However, these considerations will presumably not lead to clear-cut policy guidelines, as by their very nature, well-grounded ethical judgments have to take into account context-specific features (Prichard, 1912; Ewing, 1929; Broad, 1951; Ross, 1967 [1930]; Dancy, 2004; cf. Chapter 3). In addition, some emotional-moral concerns such as fears around nuclear energy production or genetically modified food, are more contested than others (cf. Chapter 6). In relation to context-sensitive judgments and controversial issues, a procedural approach (cf. Rawls, 1971) is best suited to take into account all possibly important ethical considerations. However, conventional procedural approaches do decision making do not explicitly address emotions. In this chapter I will develop a procedural approach for decision making about risk that explicitly encourages and includes moral emotions as important sources of ethical insight.

7.2 Emotions in Democratic Decision Making

In the previous chapters, I argued that in order to judge whether a risk is morally acceptable, more is needed than quantitative approaches, which assess risk with a function of probabilities and consequences, and cost benefit analysis. We need emotions to detect other morally relevant considerations.

However, emotions are generally excluded from political decision making (cf. Hall, 2005; Kingston and Ferry, 2008 for a critique of this). This also holds in regard to political decision making about technological risks (Sunstein, 2005, defends this; cf. Kahan, 2008, and Kahan and Slovic, 2006; Kahan et al., 2006 for a critique). This is what I previously called the 'technocratic pitfall' in Chapter 2. It means that emotions and underlying moral considerations are unduly left out of the public debate about risks.

Alternatively, some scholars argue that we should accept the emotions of the public for democratic reasons, even if they are irrational (Loewenstein et al., 2001; De Hollander and Hanemaaijer, 2003; Wolff, 2006, defend this view). For example, Loewenstein et al. argue that:

> Simply disregarding the public's fears and basing policy on the experts, however, is difficult in a democracy and ignores the real costs that fears impose on people.
>
> (Loewenstein et al., 2001, p. 281)

However, as I argued in Chapter 2, this suggests an instrumental approach, which does not take emotions seriously in their own right. It also contains the risk that emotions are manipulated without further reflection to create support or to serve a specific political agenda. For example, Sunstein criticizes policies that are based on a fear of terror (cf. Sunstein, 2005). These policies respond to people's gut reactions without critical reflection on emotions. This is what I previously called the 'populist pitfall'.

In Chapter 2, I discussed a third type of approaches, namely participatory approaches (participatory risk assessment or PRA for short). These approaches aim to include the public for democratic as well as for substantive reasons. By including the public's knowledge and values, they strive for better and more legitimate decisions. However, these approaches do not explicitly acknowledge emotions. Rather, they implicitly or explicitly demand a rational contribution from the public.

The lack of attention to emotion in PRA can be traced back to a dismissal of emotions in the underlying approaches to political philosophy, namely deliberative democracy approaches (cf. Roeser and Pesch, 2016). Arguably, democratic decision making should not just be about counting votes but should be based on a discussion about reasons, arguments and values. The importance of moral reasons in politics has been emphasized in the literature on deliberative democracy (Rawls, 1996; Habermas, 1996; Bohman and Rehg, 1997; Gutman and Thompson, 2000). However, O'Neill (2002) argues that such accounts are overly rationalistic and should include emotions as sources of moral knowledge.

Such a dismissal of emotions in political philosophy is based on an understanding of emotions as irrational or a-rational states that undermine or threaten sound decision making. According to Hoggett and Thompson (2002), approaches to deliberative democracy "lack an account of affectivity"; "either the emotions are ignored or, if they are mentioned, it is only as dangerously destabilizing forces that need to be kept in check" (Hoggett and Thompson, 2002, p. 107). However, various political theorists highlight the importance of emotions for moral knowledge, reflection and understanding, based on richer accounts of emotions (Hall, 2005; Kingston, 2011; Kingston and Ferry, 2008; Marcus, 2000, 2010; Neuman, Marcus, Crigler, and MacKuen, 2007; Nussbaum, 2013; Staiger et al., 2010).

Cheryl Hall (2005) emphasizes that paying attention to emotions in politics is important, as emotions provide us with a better understanding of the perspective of others:

> Sooner or later, that which is ignored, silenced, rejected or repressed will return. In contrast, acknowledging the dissonance, exploring it, and learning as much as possible from the different voices provides more chance of genuinely moving to a new position. [. . .] citizens who can work with their passions will be better able to develop their own perspectives as well as be more receptive to the perspective of others.
>
> (p. 130)

Hall makes an important point here: avoiding or ignoring emotions does not make them go away; rather, the concerns underlying them are bound to resurface again. Hence, it is important to explicitly address these emotions and concerns at an early stage and see them as an opportunity to learn from each other and to understand each other better.

Of course, this requires a willingness and capacity for people to deliberate with each other, and about and with one's emotions. It is a commonly accepted idea that, in order for democracy to work, people's reason needs to be educated. Drawing on work by Plato, Rousseau and contemporary feminist scholars, Hall argues that next to this, people's emotions need to be educated. She emphasizes that this does not mean manipulating people and their emotions, although she acknowledges that this is a possible danger. Rather, she argues that educating passions in the right way means enabling people to construct and change their passions or emotions. As I have argued in Chapter 6, emotions can be both the subject and the object of critical reflection. For example, Michael Lacewing (2005) elaborates on how second-order emotions, that is, how we feel about our emotions, can help us to critically reflect on and deliberate about our first-order emotions. Nussbaum (2001) argues that emotions can help us to understand the perspective of others. Hall argues that we should educate people in such a way that they develop a 'passion for democracy' (Hall, 2005, pp. 131–133). This will enable people to engage with other people's perspectives, and it will motivate them to actively participate in democratic processes.

Rebecca Kingston (2011) reasons in a similar vein, by arguing that attention to emotions is necessary for good citizenship:

> Attention to and care for the disposition and the tone through which we engage other citizens is essential for an ongoing practice of good citizenship.
>
> (p. 208)

She proposes to adopt a broader approach to political deliberation that makes room for emotions and what she calls 'public passion':

> Integrating an understanding of public passion into normative political theory will lead us to recognize a much wider sphere for political deliberation than before. This broadened sphere will include the realms of artistic expression through a number of media, and will allow for more multiple forms of exchange and debate than the traditional giving and receiving of arguments. It will involve a heightened sensitivity to the multiple manifestations of political communication.
>
> (p. 209)

Kingston's approach requires that the range of accepted forms of expression should be significantly broadened as compared to the conventional discourse that is dominated by an analytical paradigm of rationality. Conventional political discourse is mainly limited to the presentation of empirical data and deductive argumentation in which underlying values are often not explicitly acknowledged. By broadening the range of accepted forms of expression, political deliberation can more explicitly acknowledge important values and emotions, and it becomes more accessible for a wider range of audiences that might not have sufficient access to these conventionally accepted sources of knowledge and reflection.

These ideas can be extended to political decision making about risky technologies (cf. Harvey, 2009). Technological risks can have profound effects on people's well-being and values, and this should be explicitly addressed in participatory risk assessment (PRA). Science-based information and technical expertise are necessary prerequisites in decision making about technological risk, but they are not sufficient. They do not as yet provide for *moral* argumentation, which requires reflection on values, norms, virtues and ideals. This is where emotions can serve as an invaluable source of insight and deliberation in PRA. The arguments, reasons and considerations that are revealed by or lie behind emotional responses to technological risks and benefits have to be taken seriously. This makes it possible to focus the debate on important ethical issues that need to be addressed and discussed. By making emotions and underlying concerns explicit, critical reflection on whether the emotions and considerations are justified is made possible (cf. Chapter 6).

Public passion and care for others can expand beyond one's own tight community to encompass humanity. Taking on a universalistic perspective is not the prerogative of rationalism but also requires sympathy and care, as has been argued as early as the 19th century by William Whewell (1845). Indeed, people are able to care about others who live far away, as long as their destinies are presented in a way that entices emotions (cf. Slovic, 2010b, for empirical evidence). This is especially important in the context of technological risks, as they frequently transcend borders of countries, space and time.

Emotions can provide important moral insights via care, sympathy, empathy and compassion, and feelings of responsibility, justice and indignation. Emotions help us to reflect on the values that we find important and how our actions relate to our lives and those of others. Emotions also allow us to care about the well-being of others. PRA should encourage emotional deliberation about the roles that technologies can and should play, and under what conditions, so that technologies can contribute to people's well-being now and in the future, and to realizing important values such as justice, fairness, equity and sustainability.

Emotional responses to risky technologies are often especially fierce given the scientific and moral complexities involved. At the same time, emotions can make a major contribution to genuinely democratic procedures of decision making in a context where technocratic approaches are all too tempting. In order to be both effective and legitimate, participatory methods of risk assessment should recognize the full range of human reflection and deliberation, which means including emotions as an important source of moral insight.

7.3 Emotional Deliberation on Risk

Participatory risk assessment (PRA) approaches can be reformed in the light of the insights discussed in the previous section and chapters, by explicitly addressing emotions. This provides for a procedural approach to 'emotional deliberation' about risks that explicitly encourages and includes moral emotions as indispensable sources of ethical insight. I propose to revise approaches to PRA by paying attention to the following considerations.

I. Symmetrical Risk Communication

Risk communication experts often use an asymmetrical approach, which is based on one-way communication with a sender and a receiver. However, the dominant approach to communication in political philosophy is based on Habermas's (1985) '*herrschaftsfreien Diskurs*,' domination-free discourse, referring to a symmetrical, egalitarian way of exchanging ideas. The '*herrschaftsfreie Diskurs*' should also be applied to risk communication as a basis for genuinely democratic deliberation and decision making about risks. Risk communication should not only be about sending but also about receiving, listening to each other and exchanging views. Habermas' approach was grounded in a rationalistic ideal of deliberation and communication. However, by framing discussions about risk in a rationalistic way, as is usually the case, not all stakeholders are taken equally seriously. Emotionally framed concerns are easily dismissed without further discussion. My approach explicitly emphasizes the role of emotions. An emotional deliberation approach to risk invites emotions, narratives and the imagination into the arena of deliberation in order to include all relevant stakeholders and values.

II. Create Symmetrical Set-Ups of Discussions

In conventional debates about risky technologies, the hierarchy between experts and laypeople is exemplified in an asymmetrical setting where the experts are placed prominently on stage and the public forms an anonymous audience. In such a setting, it can be awkward for people to express their emotions as the experts, who usually frame their contributions in a scientific and rationalist way, dominate the discourse. Instead, all participants should be placed on an equal footing to encourage a sense of equality and empowerment for lay participants. In the case of small groups this can be done by round-table discussions. In the case of larger audiences, discussion leaders and panelists can create an interactive atmosphere by asking questions to the audience and noting their ideas on visual displays.

III. Talk about Values

It is important that values are explicitly addressed in discussions about risky technologies. However, as experts tend to dominate the discourse, it is likely that the discussion will center on scientific evidence to the exclusion of moral concerns, which are particularly vivid among laypeople (cf. Slovic, 2000, for empirical studies that show this, also cf. chapter 3). Instead, talking about values should be of utmost concern in public debates about risky technologies, as this is what the democratic decision making should be about.

IV. Talk About Emotions

Emotions are usually dismissed or not taken seriously in public debates. At the most, authorities say things like "I understand or respect your emotions, but rationally/scientifically you are not correct." Rather, one should ask what people are emotional about, especially when emotions persist in the light of scientific evidence. This might indicate that the emotional concerns are about issues other than the scientific facts. Most likely, emotions are about values. Talking about emotions can make a crucial contribution to a debate about risky technologies as they reveal values, moral reasons and considerations that can get overlooked when emotions are not explicitly addressed. By discussing the concerns underlying emotions, justified concerns can be distinguished from—morally or empirically—unjustified concerns. Taking emotions as the starting point of risk debates may reveal genuine ethical concerns that should be taken seriously. It might also show biases and irrational emotions that can be addressed by information that is presented in an emotionally accessible way. Purely rational reflection would not be able to provide us with the imaginary power that we need to envisage future scenarios, take part in other people's perspectives and evaluate their destinies.

For example, the fear of a new technology might indicate that one is not convinced of the safety measures or the supposed advantages of the technology that supposedly outweigh minor risks. Persisting emotions might also be an indication of a lack of trust in authorities. This should be addressed rather than dismissed, in order to create an open, transparent dialogue. People can perceive a dismissal of their emotions as a signal that their lack of trust might be justified. Evading an explicit discussion about a lack of trust can come across as fear and a sign that one has something to hide. Openly discussing a lack of trust instead signals respect and a willingness to engage in critical self-reflection, which is an important ingredient of genuine trustworthiness. Here lies a major role for discussion moderators: inviting people to articulate the values, reasons and considerations that underlie their emotions.

V. Ask Questions

When people respond emotionally in debates, they are often told off by debate leaders and asked to be rational. Instead, the emotional deliberation approach sees emotions as important cues to issues that matter to people which should be explicitly addressed. So rather than dismissing or avoiding emotions, debate leaders or other participants should encourage people to express the concerns underlying their emotions. Discussion leaders should invite people to tell narratives and to talk about their emotions. Asking questions can be a powerful tool. Questions can encourage people to express their concerns and values, and to critically reflect upon their own emotions and those of others, by also putting themselves in the shoes of others.

This can be done by asking people questions like:

- 'Do you find the technical information about the technology clear? Transparent? Trustworthy? If not, why not?'
- 'What are you afraid of?'

- 'What do you think could happen?'
- 'Why does that worry you?'
- 'Under what conditions would you be less worried?'

But also:

- 'Why are you enthusiastic about this technology?'

Questions could be used to engage people's imagination and reflection:

- 'Can you understand the viewpoint of the person from the other group? If not, can you try to place yourself in their shoes by listening to their story?'
- 'What problem do you think this technology has been developed for as a solution? Do you think there is a simple solution to this problem? Can we solve the problem by not using this technology? Or does that give rise to other problems? Can you try to imagine scenarios with and without this and/or other technologies? Can we solve the problems stated by designing the technology in a different way?'
- 'If you were in charge, how would you solve this problem?'

These questions can be geared toward different kinds of risks. Questions such as these allow people a genuine voice in which their emotions and concerns are appreciated, listened to and discussed. Questions such as the ones listed above can encourage people to use reflection and deliberation, not in a detached, purely analytical way, but by drawing on their emotional capacities such as imagination, compassion and empathy.

VI. Have a Dialogue Between All Involved People

In most debates on risks that involve the general public, the discussion is dominated by the experts and directed by a debate leader. However, it can be very fruitful to let people in the audience or at a round table respond to each other and to share or criticize each other's views, arguments or emotional responses. This encourages 'out of the box' thinking as a possibly surprising point of view can be explored by various participants. It can also be more forceful if people in the audience challenge each other's potentially biased opinions and emotions than if the experts or debate leaders do this. More generally, a wider dialogue contributes to a symmetrical, 'power-free' dialogue. Of course, a debate leader can intervene if people are disrespectful to each other, or if someone dominates the debate, and experts can provide scientific information if it contributes constructively to the exchange. But experts and debate leaders should dose their contributions carefully so as not to hamper an engaged dialogue.

VII. Convey Respect

The typical asymmetrical set-up of many public debates on risk signals a hierarchy, with the expert being valued more highly than the lay audience. This can convey a lack of respect. The people in the audience may lack the specific scientific expertise of the expert, but first of all, they contribute different forms of expertise based on their professions or their roles in society. Given the complexity of many decisions about risky technologies, these additional perspectives can prove valuable as they can draw attention to contextual factors that experts might not include. Furthermore, decision making about risky technologies does not only involve science but also moral norms and values. In a democratic society, all citizens should be encouraged to contribute their views, moral norms and values, and related emotions. In this case, all citizens are *prima facie* on equal footing. Hence, experts and debate leaders should communicate in a respectful, dignified way and allow an equal playing field.

VIII. Have a Clear Procedure

In order to ensure a safe and constructive setting, it should be clear to the audience what the procedure is: what happens with their input, when and how. Furthermore, the outcome of the procedure should be genuinely open. In other words, it should not be a fake consultation with preset arrangements. This was for example the case with plans concerning carbon capture and storage (CCS) in Barendrecht in the Netherlands, where the ministers told the local population at the beginning of a public debate that the decisions had already been taken (cf. Chapter 2). Arguably, the emotional response of the public was not only about the risks of CCS but about a procedure that was perceived to be unfair, undemocratic and disrespectful. Hence, rather than being a safe strategy, signaling to the public that they have no influence can backfire (Cuppen et al., 2015). An open debate, with respect for people's views and emotions, may have explicitly uncertain outcomes, but might actually be more fruitful and effective, as well as being more genuinely democratic.

IX. Appeal to People's Imagination

Other ways to involve moral emotions would be to use narratives, film and literature to make people emotionally aware of the impact of different technologies and their concomitant risks and benefits for people's lives. Several methods have been developed to enable reflection about technology, for example by letting people reflect on scenarios in which the use of a technology gives rise to moral considerations (cf. e.g., Boenink et al., 2010). These methods involve narratives that directly engage people's imaginative and empathetic capacities, and that could be further developed to explicitly encourage emotional engagement and emotional reflection.

X. Stimulate Co-Creation

Another possibility would be to let experts and laypeople co-develop scenarios for morally desirable ways in which risky technologies might be designed and play a role in society. This could be done in round-table discussions in which experts and laypeople sit together to brainstorm about possible ways to come to a more responsible innovation. Of course, in the case of a large audience, round-table discussions might not be possible, but even with a large audience, interactive, Socratic discussions are possible.

These considerations can be seen as guidelines for 'emotional deliberation' on risky technologies. Note that while several of these considerations also figure explicitly or implicitly in PRA approaches, the contribution that emotions can make is not acknowledged. However, as I argued, emotions should not be neglected or seen as 'givens' that cannot be investigated any further, as is the case with conventional technocratic or PRA approaches. Rather, emotions should be seen as triggers for discussion and as the starting point of public debate. They should play an explicit role in deliberation, by articulating and evaluating the moral considerations underlying the emotions and critically reflecting on them. Methods of participatory technology assessment can be made more fruitful by drawing on emotions and imaginative and empathetic capacities. This can empower people by letting them feel that they have something to contribute, that they are agents, not merely passive pawns in the hands of powerful companies and governments. This is especially important in the case of risky technologies, where the stakes can be extremely high.

Explicitly focusing on ethical and emotional concerns brings experts, policy makers and laypeople together on common ground, as these are capacities that all human beings share. The approach sketched here allows for a genuine dialogue where questions can be asked to everyone and where all parties should be able to listen to each other. This can stimulate people to think further, to listen to each other in a respectful way and to try to take on different points of view, thereby facilitating ethical reflection. This approach can prevent the stalemates that often occur in public debates. Public decision making about risky technologies should include the moral emotions of the public, politicians and experts. This takes us into the next section, where I will discuss the role of emotions of each of the major stakeholders in risk decision making.

7.4 Risk Emotions of Stakeholders

Decision making and debates about risky technologies involve three main groups: the public; the experts (engineers and applied scientists); and the policy makers. Conventional research on risk emotions mainly focuses on the public. In debates about risky technologies, it is the public's emotions that are often the most visible and that are contrasted with the supposedly rational stance of experts. Experts often accuse the public of being overly frightened of new technologies because it lacks the relevant knowledge and thereby bases its reactions on supposedly irrational feelings. However, as elaborated in previous chapters, the fact that the public is emotional about risks can also enable it to adopt a broader perspective on risk that can include moral considerations, which are insufficiently included in the quantitative approaches to risk that are usually employed by experts.

It might be assumed that experts take a purely rational, detached stance to risky technologies. However, scientists can be deeply emotionally involved with the research and technologies they develop (cf. McAllister, 2005). Experts and policy makers can be emotional about risks, and this can contribute to their capacity for moral reflection. Several authors emphasize that emotions are needed for moral conduct by business managers (Simon, 1987; Mumby and Putnam, 1992; Gaudine and Thorne, 2001; Klein, 2002; Lurie, 2004). We can extend this idea to other professionals and more specifically, to policy makers and engineers. If technology professionals have sufficiently developed emotional sensitivities, they will be aware of the morally important aspects of the technologies they design. Policy makers who take their own feelings of responsibility seriously will be more inclined to develop morally responsible regulations and procedures. Emotions can help all stakeholders to be aware of moral and societal responsibilities, and emotions can provide for mutual understanding.

For example, experts are more worried about nanotechnology than the public (Scheufele et al., 2007). Of course, this is partially due to the fact that most laypeople have never even heard of nanotechnology. Experts are more knowledgeable about the scientific facts than laypeople, and this can lead to their increased moral concern and worry. Apparently, the fears of nanotechnology experts can be attributed to a rational understanding of the risks involved in nanotechnology. Indeed, fear can point to a source of danger to our well-being (cf. Chapter 4, Green, 1992; Roberts, 2003). Experts should take these worries and concerns seriously, which should lead to additional precautions. Engineers can reduce the risks of a technological product by developing a different design. They have a key moral responsibility in the design process of risky technologies, as they have the technical expertise and are at the cradle of new developments. Engineers should be trained to be aware of moral values and to explicitly take them into consideration in the design process, in order to contribute to 'value sensitive design' (Friedman, 2004, van den Hoven, 2007) and 'responsible innovation' (van den Hoven, 2014). Rather than delegating moral reflection to 'moral experts', engineers should cultivate their own moral expertise, which should involve emotions. For this reason, Sunderland (2013) argues that emotions should play an important role in ethics education in engineering.[1]

Moral emotions can make experts sensitive to moral issues arising from the technologies they develop. Emotions allow us to get more deeply involved with situations. They help us transcend a detached, abstract attitude that could lead to indifference to morally problematic aspects of technologies. This is especially true in the design of risky technologies, where there might be consequences that are unforeseen or difficult to quantify, or of which we do not know whether and when they will manifest themselves. A formalistic approach to responsibility can easily lead to negligence or to the idea that 'others are responsible.' If experts are worried about the safety of the products they develop, this should be taken seriously and as a warning sign. Engineers and scientists should use their worries and concerns in the design of their research and technologies. For example, they could build barriers to prevent hazards from occurring or apply a precautionary approach, meaning that technologies of which the consequences are hard to predict should first be investigated in a safe setting. Experts should communicate their emotional-ethical concerns about technological risks and benefits to the public in addition to supplying quantitative information.

While emotions can provide each group of stakeholders with awareness of moral considerations, emotions can also make the different groups of stakeholders prone to potential biases, which can affect debates about risky technologies. We have already discussed at length the potential emotional biases of laypeople and how these can be mitigated (cf. Chapters 4 and 6 respectively). But experts and policy makers can also be biased by their emotions. For example, experts can be overly enthusiastic about their technologies, or they can be biased by self-interested concerns, such as pressure to secure funding, positions and prestige. Experts can control these potential biases by also considering themselves as part of the public[2] and trying to empathize with the point of view of potential victims of their technologies.

Risk policy makers should ideally mediate between the insights of experts and the concerns of the public. However, in practice, there can be potential conflicts of interest that might be reinforced by emotions. For example, experts from the industry lobby with regulators and often have close ties with the government when it comes to large infrastructural high-tech projects. Policy makers' careers can be at stake, which can lead to self-interested emotional biases. On the other hand, these emotions can also force politicians to follow the predominant views of the electorate. A virtuous policy maker should take a wider perspective based on feelings of responsibility and care for all members of society, and should carefully balance the risks and benefits of a technology and its concomitant moral concerns.

7.5 Conclusion

In this chapter I have developed an emotional deliberation approach to risk. This approach builds on existing participatory approaches to risk assessment or technology assessment. However, while such approaches do not pay explicit attention to emotions, my approach does so. In this way, the emotional deliberation approach can do more justice to the ambition of many participatory approaches to enhance democratic decision making and paying attention to values of all stakeholders, as emotions can provide us with special attention to what people value. By allowing all morally relevant concerns to be addressed, the new approach to risk emotions can provide for better decisions about risks. However, it can also contribute to overcoming common stalemates by offering a framework that puts participants on an equal footing; and provides for the willingness to give and take, to respect each other and to genuinely listen to each other.

Of course, the emotional responses of people can differ, but disagreement is nearly always a part of collective decision making, whether or not emotions are included. We should accept that people's emotions will probably diverge and discuss the concerns that lie behind them. Considering diverging emotions is an opportunity to develop more balanced judgments. Our emotions are not infallible. Just like other sources of knowledge, emotions can also be mistaken. As argued in Chapter 6, we should critically assess our emotions, but in doing so, we should take into account other emotions—those of ourselves and those of other people. In public debates about risk, this can contribute to an open discussion that can help overcome the stalemates that now dominate many debates about risky technologies. Taking the different perspectives of stakeholders into account can pave the way for well-grounded and well-informed policies on risky technologies. This approach can help to overcome the frequent opposition between experts and laypeople in public debates, and it can contribute to constructive solutions to the pressing ethical issues involved in decision making about risky technologies. It is reasonable to assume that there will be a greater willingness to give and take if both parties feel that they are taken seriously.

This procedure might seem costlier. However, it is likely to be more effective, and hence more fruitful in the long run. Currently, many debates about risky technologies result in an even wider gap between proponents and opponents and in rejections of technologies that could contribute positively to society if developed and introduced in a morally sound way. Genuinely including emotional concerns in debates about risky technologies can help overcome such predictable stalemates. To illustrate this, in the next and final chapter, I will discuss several salient cases that show the potential applications of my approach in ethical and political deliberation about risky technologies.

Notes

1. The 'rationalistic' bias in current engineering culture is also reflected by the fact that engineering is regarded as a 'male' profession with a low percentage of female engineers and engineering students. This is largely because the concepts 'rational' and 'male,' and 'emotional' and 'female' are traditionally linked (cf. Faulkner, 2000; Robinson and Mcllwee, 2005). Turning engineering into a 'softer' discipline might also have an effect on gender roles, possibly making engineering a more attractive discipline for women.
2. Van der Burg and Van Gorp (2005) make this point based on an argument from virtue-ethics.

8 Emotional Deliberation on Technological Risks in Practice

8.1 Introduction

In the previous chapters, I developed a new theory of risk emotions that sees them as sources of practical rationality, and that proposes to take emotions as the starting point of debates about risky technologies. In this chapter, I will discuss a few cases that illustrate the contribution that my new theory can make to the public debate on, namely, nuclear energy, climate change, public health risks and architecture. These are important societal issues that give rise to heated debates and stalemates. In all these cases, my approach to risk emotions can provide a new perspective by laying the foundation for a more constructive dialogue and public deliberation. In addition, the case of climate change also illustrates the potential role of risk emotions in fostering morally responsible action, even when it involves personal sacrifice. The case of architecture also points to an additional evaluative dimension of risk, namely aesthetics. The chapter concludes with further expanding on the theme of aesthetics by exploring the contribution that technology-related art can make to emotional deliberation on ethical aspects of technological risks.

8.2 Moral Emotions and Risky Technologies: Nuclear Energy

On March 11, 2011, an earthquake and tsunami hit the coast of Japan, giving rise to a nuclear meltdown in the Fukushima Daiichi power plant. This resulted in a new debate about nuclear energy. In the years preceding the accident, there was a growing consensus that in order to decrease CO_2 emissions, nuclear energy should play an important role in generating energy. The probability of an accident was said to be negligible. However, after the Fukushima accident, nuclear energy has become controversial again and people have argued that we should abandon it (cf. e.g., Macilwain, 2011). Germany immediately shut down several nuclear reactors, and because of its anti-nuclear position, the German Green Party achieved unprecedented results in local elections.

Despite this shift in focus, there seems to be one constant factor in the debate about nuclear energy: proponents call opponents badly informed, emotional and irrational, using these notions more or less as synonyms. This type of rhetoric is denigrating and hinders a real debate about nuclear energy. In addition, it is simply wrong to equate emotions with irrationality, as they can be a source of practical rationality, as I argued in the previous chapters. The stereotypical rhetoric according to which experts are rational and objective on the one hand, and the public is emotional and irrational on the other, should be avoided. The picture painted is empirically false and prevents a fruitful debate from taking place. The approach developed in this book indicates that a fruitful debate about nuclear energy should do justice both to quantitative, empirical information as well as emotional, moral considerations. Although part of the required information, namely the quantitative data, does lie with experts, the experts do not have privileged access to the moral considerations that are necessary for assessing the acceptability of a risky technology such as nuclear energy. Moral emotions such as imagination, sympathy and compassion can provide us with important perspectives. For example, we only start to grasp the moral impact of the disaster in Fukushima if we see pictures and hear the stories of the people who were evacuated, small children who were tested for radiation contamination and safety workers who were taken to hospital with burns, and if we think about the uncertainty concerning the consequences that people experienced in the direct aftermath of the accident and even years ahead. The moral meaning of a disaster like this only starts to become clear if we are emotionally engaged with the people who have to undergo the consequences.

Nuclear energy's most salient risk is that of a meltdown. It is often argued that the chance of a meltdown is low, but how low is highly debated (cf. several contributions to Taebi and Roeser, 2015). In any case, meltdowns can lead to large-scale consequences, and that prospect gives rise to intense public concern, especially in the direct aftermath of a nuclear accident such as at Chernobyl or at the Fukushima Daiichi power plant.

The probability of a nuclear disaster might be small, but if it occurs, the consequences are enormous. In addition to the supposedly low probabilities of a nuclear disaster, it is also important to focus on its possible consequences. Is the meltdown of a modern reactor less disastrous than a meltdown in an old one as in the case of the reactor in Chernobyl? How reliable are the safety barriers? The other side of the coin is that we should also not forget the hazards involved in mining and using coal, and the general environmental and health effects of CO2 emissions. Are there acceptable alternatives for nuclear energy if people are unwilling to reduce their energy consumption and oppose large-scale wind turbine parks as well? These considerations illustrate the complexity and intricacies of the technical and moral considerations involved in contemplating the moral acceptability of nuclear energy.

Nuclear energy might be unavoidable given our energy consumption, which in turn means that we have to find a solution for nuclear waste. Even if we would decide to stop nuclear energy now, we still have the problem of the nuclear waste resulting from previous decades of nuclear energy production.

On top of the waste generated at nuclear power plants, we also have to find solutions for the radioactive waste from medical applications. These are generally not considered controversial, presumably because people recognize the usefulness and irreplaceability of the applications. Furthermore, in many parts of the world there is an increase in nuclear energy facilities. In other words, nuclear waste is here to stay. Given the time-scale and geographical aspects of nuclear waste, international solutions for nuclear risk management and nuclear waste management may be unavoidable (Taebi, 2012b). There is the temporal dimension of nuclear waste, which is tremendously long. It gives rise to intricate moral and practical issues, such as intergenerational justice (cf. Taebi, 2012a): we burden future generations with the waste of our activities. Furthermore, this involves the question as to how to communicate with future generations, in which language and through which medium; whether they should be warned about the dangers of nuclear waste, or whether they should not be informed as to its location in order to prevent abuse.

The emotions of opponents of nuclear energy include the fear of a catastrophic event. The emotions of proponents include curiosity and enthusiasm for the potential and benefits that nuclear energy might offer compared to, for example, coal generated energy. These emotions reveal important evaluative aspects of nuclear energy and should be taken seriously in the debate. By addressing opposing emotions as starting points of debates rather than ignoring them or taking them as endpoints of debates as is currently often the case, the underlying ethical concerns can be made explicit and discussed. A fruitful debate about nuclear energy should do justice to quantitative, empirical information as well as to emotional, moral considerations, as I already indicated in discussions in previous chapters (also cf. Roeser, 2011b; Taebi et al., 2012; Nihlén Fahlquist and Roeser, 2015, for further discussions of emotions and values in the context of nuclear energy; cf. Taebi and Roeser, 2015, for discussions of ethical aspects of nuclear energy from a variety of perspectives).

8.3 Moral Emotions and Systemic Risks: Climate Change

Climate change is one of the major challenges facing the world in the 21st century and beyond. Climate change is an extremely urgent problem that presumably will affect the environment for generations to come, and it will also have effects on the health and way of life of present and future generations (cf. Moser, 2010; Hulme, 2009, for reviews of the literature on climate change). There is a growing consensus that climate change is caused by human activities, but few people are willing to significantly adapt their lifestyle in order to reduce their ecological footprint. Sheppard (2005, p. 652) states, "There is an alarming gap between awareness and action on climate change." Several researchers who study the perceptions that people have of climate change have stated that people lack a sense of urgency (Leiserowitz,

2006, p. 64; Lorenzoni and Pidgeon, 2006). A recurring theme in explanations of this lack of urgency is the lack of personal, emotional involvement with the possible effects of climate change (e.g., cf. Lorenzoni et al., 2007; Leiserowitz, 2005, p. 1438).

Lorenzoni and Pidgeon (2006, p. 85) argue that people feel that their own contribution would be futile; people demand action at a collective policy level instead. However, we should note that, at least in democratic societies, public policies are driven by the preferences and demands of citizens. It could be the easy way out for people who claim that the government should take care of policies to address climate change. It is arguably necessary that citizens put this topic on the political agenda and show that they are supportive of policies that might require sacrifices from the public. Yet, it is doubtful whether there currently is indeed such support. Leiserowitz found that "the public largely supported policy action at the national and international levels, but opposed two tax policies that would directly affect them". (Leiserowitz, 2006, p. 56) Leiserowitz (2005, 2006) and Meijnders et al. (2001) explain this reluctance to make sacrifices by the fact that people perceive the risks of climate change to be remote:

> The farther away in time and space people think a threat is, and the more difficult it is for them to visualize the threat, the less involved they are . . . This nicely captures one of the key challenges in climate change policy: How to legitimize drastic policy measures against a problem as "far away" and as "abstract" as climate change.
>
> (Meijnders et al., 2001, p. 965)

Several scholars have proposed that emotions might play a role in risk communication strategies targeted at adapting lifestyles to diminish climate change (e.g., Weber, 2006). Similarly, Meijnders et al. (2001) argue that communication about climate change should appeal more directly to feelings such as fear. The approach to risk emotions developed in this book lends additional strength to these claims by empirical climate risk communication scholars. Emotions might be the missing link in effective communication about climate change. They have a two-fold role: they lead us to greater awareness of the problems; and they motivate us to do something about climate change. I will explore these two roles of emotions in what follows.

The main goal of risk communication is to inform the public about risks. However, in the case of climate change, this gives rise to several problems. The first problem is that the scientific explanation of climate change is complex and not uncontroversial (although cf. Moser, 2010, for a discussion of how the idea that the evidence is not uncontroversial might be an artifact of biased communication and framing through the media). The question arises whether it is justified to take action on the grounds of insecure knowledge. The precautionary principle addresses this point, taking the stance that it is 'better to be safe than sorry', hence to take precautionary measures despite incomplete scientific evidence (cf. Ahteensuu and Sandin, 2012). Here, emotions such as worry and care can play an important role by letting us take on responsibility for our actions and making personal sacrifices, even though our contribution might be futile or insecure.

An additional problem is the possibly limited knowledge of consumers. Even if people have the goodwill and commitment to lead a sustainable lifestyle, they often have limited knowledge of how products are produced. This points to the responsibility of producers to provide information about production processes, and this probably has to be enforced through legislation, on top of legislation requiring sustainable production methods. Obviously, not only do consumers have the moral obligation to contribute to a sustainable world, but industry and politics have this obligation as well. These different actors are dependent on each other and can easily pass the buck around, with the risk that nothing happens. However, here again emotions of compassion with potential victims and feelings of responsibility can provide a way to transcend this vicious circle. Risk communication about climate change should be directed at the emotions of politicians and people working in industry as well as at the emotions of the public.

An additional problem is that the way risks are presented determines, to a large degree, how information is understood by the recipient. This can lead to distortions, misunderstandings and biases ('framing': Tversky and Kahneman, 1974; cf. the discussions in chapters 3 and 4). Gigerenzer (2002) argues that this phenomenon undermines informed consent—that is, the idea that autonomous agents can make decisions about the acceptability of risks based on sound information. Intuitions and emotions are generally taken to be unreliable sources of insights into the quantitative aspects of risk (Gilovich et al., 2002; de Sousa, 2010). However, as I have argued in this book, intuitions and emotions are indispensable in discerning ethical aspects of risks. In risk communication, emotions should not be abused for manipulative purposes; rather, they should be seriously addressed in order to trigger reflection (similar ideas have been proposed by Sandman, 1989, and Buck and Davis, 2010).

Several authors who argue for a more important role of emotions in communication about climate change nevertheless mention that we should be aware about possible ethical problems with this approach, as it might lead to the manipulation of the public (Meijnders et al., 2001, p. 965; Sheppard, 2005). One might argue that as long as it is for a greater good (in this case to mitigate climate change and its devastating consequences), manipulation is justified. However, this is a consequentialist way of reasoning that is ethically problematic, as it might not respect the autonomy and reflective capacities of people. Thaler and Sunstein (2008) argue that manipulation is unavoidable; no matter how options are presented, they frame our choices and behavior. They argue that given this fact, we should provide choice options ('nudges') that steer us in the directions that we would endorse. However, given my proposed view of emotions, emotions are not simply manipulative measures.

Instead, emotions can enable moral reflection and deliberation. Indeed, Meijnders et al. (2001, p. 965) argue that 'fear appeals may stimulate people to think and to be critical decision makers.' The approach I have developed in this book supports this idea. Rather than being a form of manipulation or nudge, appealing to moral emotions about climate change can enable more thorough ethical reflection about the impact of climate change.

This means that emotional appeals should not be limited to alarmist images, but should also provide narratives and portraits of people who are affected by climate change, and who may not lead a polluting lifestyle themselves. This enables critical reflection about one's own lifestyle and considerations of justice toward others. By providing people with concrete narratives, distant others who can otherwise be easily neglected come uncomfortably close and force oneself to critically assess one's own behavior (cf. Spence and Pidgeon, 2010, for empirical findings that confirm this). Communication about climate change should appeal to these reflective moral emotions as they give rise to critical ethical reflection.

So far, I have argued why emotions are needed to fully grasp the moral meaning of climate change. However, emotions are generally considered to be intrinsically motivating states (Scherer, 1984; Frijda, 1986, p. 77; Ben-Ze'ev, 2000). This means that integrating emotions in the debate about climate change can serve two purposes: one, it can lead to a more thorough understanding of the moral impact of climate change by sympathizing with its victims and future generations, as argued above; but at the same time, two, it can serve as a more reliable source of motivation than purely rational, abstract knowledge about climate change.

Ideas by the philosopher Linda Zagzebski about the relationship between emotional moral judgments and motivation can help us to understand how emotional moral judgments about climate change can lead to changed motivation and behavior. According to Zagzebski, emotions are unitary states that have a cognitive and an affective aspect (Zagzebski, 2003, p. 109). She argues that in moral judgments, cognition and affect can go together but they need not necessarily do so. This is what Zagzebski calls the 'thinning' of moral judgments, which means that they can become less emotional and hence less motivating. She says that 'ground level moral judgments' are the most basic moral experiences on which our more abstract moral judgments are based. A ground-level moral judgment is directed toward a concrete case here and now. When we reflect upon a situation, we abstract from a concrete experience, and this weakens the motivating emotion that was there in the initial judgment. The most abstract and usually least motivating moral judgments concern general moral principles.[1] This means that if the feeling aspect of emotions becomes less intense, the motivating aspect becomes less intense as well.[2]

Emotional reactions to directly observed morally reprehensible behavior by others typically gives rise to a desire to intervene. This is less so in the case of more detached, abstract moral judgments, for example after reading a newspaper headline. I also discussed this in Chapter 5 regarding Paul Slovic's empirical findings concerning decreased compassion and moral action in the case of increased numbers and abstraction (cf. Slovic, 2007). In a similar way, statistical information about climate change can be shrugged away easily as it is abstract and not attached to meaning. Based on Slovic's work, we can say that this can be overcome by presenting information in a way that appeals to emotions such as feelings of justice and sympathy for victims of climate change, in present and future generations. Complex statistics can be replaced or supplemented by understandable, gripping narratives (cf. Moser, 2010, p. 36) and by involving the arts (Moser, 2010, p. 43). This resonates well with the work by Nussbaum (2001), who emphasizes the role of art and narratives to expand our capacity to feel compassion from those who are close by to more distant others. Increased, more intensely felt compassion can then provide for stronger motivation to act accordingly, even if it means that we have to make personal sacrifices, such as adjusting our lifestyle to minimize our ecological footprint.

A potential problem is that apart from having a positive impact on our behavior, emotions can also have negative effects, for example in the case of fear. Fear can be a motivating factor. It could, for example, cause us to avoid a fearful object or change our behavior, but it can also be paralyzing. Roser-Renouf and Maibach (2010) point out that using fear messages can be problematic as they might emphasize the futility of our efforts. How we respond to fear depends on personal and other circumstances (cf. Roeser, 2011a, p. 175; also cf. Rothman and Salovey, 1997, on the complexity of enticing behavior through specific frames). Other emotions can help to balance paralyzing emotions (Roeser, 2011a, p. 176). In a similar vein, Roser-Renof and Maibach (2010) argue that in addition to fear, what is needed is hope, for example through vivid concrete examples. Hulme (2009) makes a similar point by arguing that we should not see climate change as an insurmountable problem, but rather as a challenge and as a source of imagination about our social responsibilities.

Hence, in order to let our moral judgments result in motivational states, we should involve our emotional capacities. As I argued in this section, this means that deliberation and communication about climate change should integrate moral emotions for two reasons: one, because moral emotions lead us to more substantiated moral insights about climate change; and two, because they provide the motivation to adapt our behavior. Emotions give us a more substantial grasp of the meaning of morally relevant considerations about climate change, and they provide the motivation to behave accordingly. In the case of climate change, emotional considerations might be the key to changing our behavior effectively. Communication about climate change should trigger moral emotions to entice moral reflection and motivation for a more sustainable lifestyle. These insights can be applied to other debates about risk where people find it difficult to adjust their behavior (cf. Roeser and Nihlén Fahlquist, 2014).

8.4 Emotions, Health Technologies and (Public) Health Risks

Technologies also play an important role in medical developments. This is an area where biomedical ethics and technology ethics can shed more light on each other. Indeed, as I will argue, the emotional deliberation approach to risk can provide fruitful insights into highly sensitive issues of ethical and risky aspects of health technologies.

In the context of health risks, there is often a tension between public and individual health and well-being. Public health policies are targeted at overall, aggregate improvements in levels of health and well-being. These policies are often based on statistical and cost-benefit, consequentialist approaches, similar to the technocratic approaches that I criticized in previous chapters. In these chapters, I argued that these approaches fail to include important ethical considerations such as justice, equity, fairness, autonomy, available alternatives and context-specific circumstances. In the context of health, moral emotions can also provide for important insights into these ethical considerations, for example by feelings of responsibility and fairness, and by engaging with compassion and care with individual's specific circumstances. The moral considerations mentioned play an important role in standard approaches to biomedical ethics (Beauchamp and Childress, 2006), and moral emotions also play an important role in the ethics of care (Held, 2006). These approaches are primarily directed at contexts where individual interactions between health care providers and patients are central. However, in the often-technocratic world of public health and health technologies, these moral considerations and emotions do not figure prominently. In the remainder of this section, I will discuss a few examples of medical technologies that could be improved by employing the emotional deliberation approach developed in this book.

Infant Feeding, Risk and Emotion

'Breastfeeding is best for your baby'—this slogan is heard by all prospective parents from the beginning of the pregnancy to childbirth and in the months after that. Breastfeeding is the most natural food for babies and is praised for many health advantages over formula milk, which is considered artificial and a technological product that is unnecessary. Numerous scientific studies show the health benefits of breastfeeding, mostly for the baby but also for the mother. Based on this information, the WHO, UNICEF, national governments and health organizations promote breastfeeding. Hospitals that encourage breastfeeding can get a specific certificate.

Many young parents indeed want to breastfeed their babies but are not always successful. The international nursing organization La Leche League disseminates information about breastfeeding. Lactation consultants are available to assist with breastfeeding. However, despite all this support, there are still situations where breastfeeding does not work. In the Netherlands, for example, the vast majority of mothers breastfeed their baby after they have given birth, but after just one month, that number is halved, and few mothers breastfeed their babies up to six months after birth, even though this is the recommended minimum duration (Lanting and van Wouwe, 2007).

There are several difficulties associated with breastfeeding, including retracted nipples; mastitis (which makes feeding very painful); early or difficult births; physiological issues making it difficult for the baby to suck properly; psychological problems so that the mother has difficulty feeding; insufficient milk production; and difficulties with combining breastfeeding with other tasks, such as work or the care of other children. While many women experience problems, the official bodies repeatedly state that 'everyone can breastfeed' and that the problems mentioned above can be overcome with the right support.

This means that the burden of proof for unsuccessful breastfeeding lies with the mothers: apparently they have not done their best. But this is not always the case. On the contrary, many young mothers continue trying to breastfeed, despite problems, until they are completely exhausted or perhaps even worse given the lack of choice, until the baby is exhausted and malnourished. In these cases, care providers such as midwives, lactation consultants and doctors frequently respond with the advice to keep trying, arguing that breastfeeding is best for the baby.

The question in this case is whether a well-intentioned measure, the international policy to encourage breastfeeding, has not been taken out of context and does more harm than good. The breastfeeding policy was initially intended for women in developing countries who have no access to clean drinking water and affordable formula, which means that giving formula milk could pose a health risk. But in Western countries bottle-feeding is a viable alternative in cases where breastfeeding fails. There are even studies showing that bottle-feeding is equivalent or superior to breastmilk vis-à-vis the health of babies (cf. Wolf, 2011 for a review of the conflicting evidence), but these studies are almost never mentioned. This means that the information is lopsided. And while the information might be intended for women who would otherwise not consider breastfeeding, educated, healthy and environmentally conscious women in particular are eager to breastfeed. Breastfeeding is an important part of their ideal of what it means to be a good mother, and it fits their overall worldview and their cherished values. Despite this, they too frequently encounter problems with breastfeeding. Not being able to breastfeed is therefore often particularly painful for these women, because it challenges their identity as a mother and as a person (cf. Nihlén Fahlquist, 2016).

Furthermore, after giving birth, women are already more vulnerable given the new situation in which they find themselves, let alone having to deal with the hormonal, physical and mental preparations for breastfeeding. What these women need is understanding and support should they decide to switch to bottle-feeding. Unfortunately, in practice it is usually the opposite: health care providers continue to stress that the mothers should continue with breastfeeding and this frequently results in feelings of frustration and guilt (cf. Nihlén Fahlquist, 2016).

The policy of promoting breastfeeding is a case in point in the limitations of an approach to health risks that is primarily based on statistical evidence without paying proper attention to the contextual and qualitative aspects of risk that I discussed in previous chapters. The policy does not look at aspects such as feasibility, available alternatives, justice, care for and autonomy of young parents (for a more detailed discussion of these aspects in the context of breastfeeding, cf. Nihlén Fahlquist and Roeser, 2011). An alternative policy would be based on greater understanding and more realistic information on breastfeeding, and it would take into account the emotional impact that information can have on people in vulnerable situations. Emotions such as care, sympathy and compassion can help health care providers as well as policy makers to provide more nuanced and personalized information on this matter, which has huge emotional impact for many people. Emotions of care, sympathy and compassion by health care providers could contribute to a more understanding and helpful attitude that appreciates the efforts and sacrifices that parents have frequently already made, and it may support these parents through the process of making a decision that they find very difficult—that is, to stop breastfeeding.

HPV Vaccinations

In recent years, it has become possible to vaccinate teenage girls against the human papilloma virus (HPV), which is the cause of certain forms of cervical cancer. These vaccinations turned out to be extremely controversial. In the United States the opposition was primarily related to norms concerning sexual behavior, arguing that chastity is a more appropriate protection against HPV infection (cf. Haber et al., 2007 for a review of the controversy). In the Netherlands, the opposition was based on a very different concern. There was supposed evidence that girls had severe side effects after HPV vaccinations, such as narcolepsy and even death. A network of 'concerned mothers' started a vociferous campaign against HPV vaccinations and spread supposedly scientific information on incidents. This led to profound concerns and a refusal by parts of the population to vaccinate their daughters.[3] The Dutch health authorities responded to this by assuring them that the vaccinations were safe and the health risks were negligible. This information failed to convince concerned parents.

The emotional deliberation approach can shed light on this case. The Dutch health authorities relied on a scientific, technocratic approach while the parents were concerned about the health effects of the vaccination on their daughters. If the health authorities had explicitly adopted a compassionate perspective, they might have understood the parents' concerns better. Instead, the message that the parents heard was: 'don't worry, the risk that your beloved daughter will die from this is very small.' It is only natural for parents to want to avoid any risk to their children whatsoever, be it death or health defects. If the health authorities really think that the evidence supports HPV vaccinations for all teenage girls, they should not primarily focus on the low risk of side effects, but they should initially discuss their main motivation for promoting HPV vaccinations, namely trying to prevent girls from developing a terrible disease in later life. This concern actually directly parallels the parents' concerns: they are wary of the HPV vaccination because they do not know how it will affect the well-being of their daughters. The health authorities and the parents should thus be able to find common ground. Presumably, the health authorities were hesitant to use the argument of the well-being of girls in order to avoid being seen as wanting to manipulate parents by appealing to their emotions. However, while the conventional framework to risk deems emotions to be by definition irrational and a threat to sound decision making, the emotional deliberation approach would offer an alternative perspective. Engaging with emotions can provide for a more nuanced outlook. Information on both the positive and negative aspects of vaccination could be provided in a balanced and nuanced way, highlighting the complex scientific as well as the moral and emotional aspects. Solely focusing on the scientific aspects and reducing them to statistical numbers actually means withholding a complete, nuanced outlook on this complex matter from parents. This in turn means that the parents are not optimally enabled to make decisions concerning the possible vaccination of their daughters. So in this case, avoiding emotions actually means providing an incomplete basis for decision making.

Emotional Deliberation on Online Fora on IVF Treatment

The Internet provides people who suffer from similar health problems with numerous platforms to exchange experiences. In an online study of fora for women with fertility problems undergoing IVF treatment, Sofia Kaliarnta (2016) has found that these women exchange their experiences, emotions and concerns in a way that they themselves state they cannot find elsewhere. The fora constitute genuine friendships and support. Neither the families of these women nor their health care providers are able to understand their experiences as well as other women who are in the same situation. The anonymity of the fora as well as their shared experience enable them to engage in-depth with each other and provide each other with care and support. In this case, the fora can be understood as a technology that enables women to emotionally deliberate on their experiences and concerns with another technology, that is, IVF treatment (Kaliarnta et al., 2011; also cf. Veen et al., 2010 concerning online fora of patients with coeliac disease).

Fear and Hope Concerning GM Virus Treatments

A recent development in cancer treatment is to use genetically modified (GM) viruses that can attack tumor tissue and provide treatments against cancer. These treatments are still at a fairly experimental stage. They require patients to stay in isolated hospital rooms in order to avoid shedding the GM viruses into the environment, which might lead to unforeseeable environmental and health risks. While these treatments can have severe side effects, patients are frequently willing to undergo these treatments (cf. Mampuys and Roeser, 2011). This is interesting as genetic modification is very controversial when used in agriculture, but apparently less so when applied to health contexts. This can be partially explained by the taxonomy that I provided in Chapter 6, where I argued that people see technological interventions as less controversial in health contexts despite their risks. My explanation for this phenomenon was that illness is perceived as a deviation from the healthy status quo that can be restored with a health technology. This is in direct contrast to the perception of technology in other contexts, such as energy or agriculture, where technologies are aimed at improving well-being from a status quo that is often already perceived as acceptable. A parallel situation can be seen in the context of radioactivity. Radiotherapy for cancer patients is common and uncontroversial because it is perceived as potentially life-saving in a context where there are no other real alternatives. However, nuclear energy is considered to be controversial, unnecessary and easily avoidable in the face of alternative energy sources.

This general willingness to accept medical treatments, even when they are still in an experimental stage and the success of the treatment is as yet unclear, can mean that people might not be fully aware of the impact these treatments can have on their well-being. For example, dosages in the experimental stage tend to be low and might not even be fully effective, but they may bring with them profound negative side-effects and require quarantining the patient to avoid shedding (cf. Mampuys and Roeser, 2011). Patients are reported to be aware that the treatment might not help them but may help future patients, and they are willing to undergo negative side-effects for that reason. This can provide them with a meaningful experience, namely that their suffering is not in vain, even if they themselves might not benefit from it (cf. Mampuys and Roeser, 2011). The emotional deliberation approach can provide for important nuances by discussing the impact the treatment might have on patients' well-being in a compassionate way. Patients would thus be aware of the expected trajectory and have genuine opportunities to reflect on them, with the consequence that they may even change their minds about participating. By paying attention to the relevant aspects of the situation and to the patients' experience, their consent may be more informed.

These examples from health technologies illustrate how the emotional deliberation approach can provide important insights in ethical aspects of risk, as well as pay attention and show concern for individual well-being in contexts where it might be tempting to focus on aggregate public health goals that lose sight of individual experiences.

8.5 Moral Emotions and Aesthetic Risks: Architecture and the Built Environment

The previous examples discussed in this chapter have dealt with specific, controversial technologies, systemic risks and risks of health technologies respectively. Common factors in my discussion of these technologies were a focus on the impacts on physical and emotional well-being, albeit in different ways, and the highlighting of additional qualitative aspects of these impacts other than those commonly considered in the prevailing quantitative, technocratic approaches. These additional qualitative aspects also rarely receive sufficient serious attention in more populist approaches as well as in participatory approaches that may not go far enough in engaging with people's deeper concerns.

I will now look at a technology that gives rise to yet another qualitative dimension: architecture (and urban planning). When it comes to risky technologies, architecture is a special case. Most technical artefacts can create risks for health and the environment. However, in the case of architecture, there is an additional dimension involved—that is, an *aesthetic* dimension. I will argue that a qualitative approach to risk should supplement ethical values with aesthetic values. Architecture creates artefacts that influence our *visual* environment.[4] A building can be an eyesore. Venturi et al. use the notion of 'visual pollution' for this phenomenon (Venturi et al., 1972). In contrast, other buildings are beautiful and contribute positively to the aesthetic quality of a city or town. Buildings matter aesthetically. But this fact also influences an area's quality of life and well-being. Here, ethical and aesthetic concerns coincide (Harries, 1997; Taylor, 2000). Hence, we can say that buildings should not only be sustainable from the point of view of health and the environment, but also from an *aesthetic* point of view.

The importance of aesthetics for well-being is widely acknowledged in architecture theory, and many scholars and architects are aware of the moral responsibility that architects have in designing buildings that make a positive aesthetic contribution to society. However, in the literature on technological risk, aesthetics has not yet been acknowledged as a factor that poses threats or risks to people's well-being. In other words, aesthetics has not yet been conceptualized as a risk factor. However, this conceptualization may lead to fruitful insights. This section aims to connect the discussion on the moral responsibility of architects to design aesthetically acceptable buildings to the discussion in risk theory in which qualitative risk factors are included in the conception of morally acceptable risks. The proposal made in this section is to include aesthetics as a qualitative risk factor in the discourse on risk-ethics, as it makes an important contribution to the existing lists of qualitative risk factors.

Here are a few dimensions along which aesthetics fits within the risk-ethics framework. One, as stated above, bad aesthetics can affect our well-being, which is where ethics and aesthetics intersect. Two, aesthetics comprises uncertainty: how will people appreciate the aesthetics of a building a few decades from now? Three, bad aesthetics threatens future generations by burdening them with potentially horrendous buildings. Four, aesthetic risks give rise to issues of justice, fairness and autonomy. For example, wealthy people have more freedom and power to choose the aesthetic environment in which to live, just as they have more freedom and power than poor people to choose healthy and sustainable environments to live in. In as much as aesthetics contributes to people's quality of life and well-being, injustices concerning other aspects of quality of life and well-being also play a role in the context of aesthetics.

As discussed in the previous chapters, the conventional, technocratic approach defines risk as a probability of an unwanted effect, and then applies cost benefit analysis to determine the risky activity that has the lowest net risk. Ethical considerations such as justice, fairness and autonomy put boundaries on the cost-benefit analysis, mirroring the deontological and virtue ethical objections against consequentialist approaches in ethics. But ethical considerations also play a role in the determination of the kinds of effects to consider in a risk assessment, such as whether to look at annual fatalities or also include sick or injured people, or effects on nature. The proposal to take aesthetic considerations in risk assessment into account also concerns the kinds of effects to take into account. It broadens the scope of morally relevant consequences in risk assessment from matters of life, death, health and physical well-being to aesthetic well-being.

Authors who write on risk-management, risk perception and risk-ethics have largely neglected the aesthetic dimension of technologies.[5] This is a major omission. Although there is a research community on environmental aesthetics (cf. Carlson, 2007, for an overview), its discourse exists separately from the risk-ethics discourse. It would be fruitful to bring these two discourses together as they would both benefit from their respective insights and expertise. Specifically, it would enable risk-scholars to gain a more complete account of ethically relevant considerations in thinking about risky technologies.

Connecting the two debates of aesthetic responsibility of architects and risk ethics is more than a merely theoretical exercise. As these two theoretical approaches have direct practical implications for professionals and policy makers, connecting them will make their practical relevance even more explicit. More specifically, by leaving aesthetics out of the current debate about acceptable risk, there is a danger that aesthetics will be left out of the decision-making procedure altogether. It is important to include aesthetic considerations with other morally relevant considerations from the outset in order to proactively reflect on how to do justice to all morally relevant risk considerations, including aesthetic considerations. By leaving them out of the 'risk equation', there is the danger that qualitative risk aspects will only emerge as an afterthought. It is much more effective to take qualitative risk considerations such as justice, fairness, autonomy, equity and aesthetics into account from the start. This will enable the best solutions to be found that do justice to them all in the most morally responsible way.

Taking aesthetics in risky technologies into account allows a broader range of morally relevant features in risk assessment to be considered. These ideas can be extended to urban planning and to aesthetic aspects of nature that might be threatened by human activities. Here are several examples. When urban planners design a highway, they have to take into account the level of noise and the quantity of emissions that might affect the health of people living in the vicinity; but they also have to take into account the visual effect that the highway has on the landscape. A highway is more controversial when it cuts through a natural landscape than when it cuts through a built-up area. Another example is wind turbines and solar cells. Wind energy and solar energy are more sustainable than other sources of energy in terms of CO2 emissions. However, critics of these sources of energy point to the fact that massive wind turbine or solar farms can harm the *visual* aspects of our environment. This is an important consideration that has to be given due weight. This does not imply abandoning windmills and solar cells because of their potential to harm the environment aesthetically; rather, in planning appropriate locations for these technologies, they should be given sufficient consideration. Options could be placing wind turbines along highways or in harbors rather than in the middle of rolling hills as these are already visibly human-made, industrialized environments. In the case of solar cells placed on roofs of buildings, the solar cells could be designed in such a way that they do not detract from the design of the building. They should either be as invisible as possible or become a part of the design of a building.

These examples illustrate that aesthetic aspects already play an important role in debates about environmental issues. However, greater conceptual clarity can be achieved by explicitly including aesthetic aspects in the thinking about the ethics of risky technologies. If aesthetic considerations are not part of the conceptual framework of risk assessment, there is a danger that they are left out as being supposedly irrelevant. The methodologies for assessing risky technologies have to be formulated in such a way that all morally relevant considerations are given due weight. This means that aesthetics should be added to the ethics of risk as yet another qualitative risk factor that we have to explicitly take into account.

In the case of including aesthetic considerations in a risk assessment, the fundamental question is how to trade off aesthetic considerations with considerations about the physical well-being of humans and nature. As argued above (cf. Chapter 2), while theoretically it is not impossible to design a quantitative model that makes trade-offs between different kinds of values, the philosophical question remains whether these are morally justified trade-offs and whether the same trade-offs can be made in each and every case. Hence, broadening the scope of risk assessment of architecture (and urban planning) to include aesthetic considerations will inevitably involve contextual, situation-specific deliberation. Here the approach developed in the previous chapters can make an important contribution, but it should be broadened to include aesthetic emotions and values next to moral emotions and values. I will argue that deliberation on aesthetic risks of architectures should involve different stakeholders, including architects, the public and policy makers, as they can provide for a broad range of insights, based on moral as well as aesthetic emotions.

How should architects and urban planners deal with the proposed insight that aesthetics is a risk factor? The challenge for architects is to take a timeless aesthetic taste into account. They should strive to design buildings that are not only sustainable from an environmental and health point of view, but also from an aesthetic point of view. They should design buildings and urban settings that do not just live up to the latest fashion, but that will presumably also be appreciated by people who live in the future.

From a pessimistic viewpoint, one might think that it is impossible to predict how people will perceive buildings in the future. Postmodern or social-constructivist approaches to aesthetics deem that there is nothing like objective aesthetic criteria; aesthetics is merely a social construction, and it is hard to predict what people will appreciate in the future. A sceptic's approach may be that even though there might be objective aesthetic criteria, people might just not grasp them. In these views, aesthetics might be more a matter of 'ignorance' than risk. In risk literature, authors use the notion of ignorance for future effects of technologies that are beyond prediction.

However, these views are too pessimistic. There are aesthetic theories that propose that aesthetic values are objective and that people can have access to these aesthetic values. Carlson (2000) argues for an objectivist account of environmental aesthetics. This view is analogous to the theory of realism in metaethics, according to which moral values are objective and people can have knowledge of them (cf. Shafer-Landau, 2003; Cuneo, 2007; Audi, 2003; Roeser, 2011a; and Enoch, 2011). Note that from an objectivist, realist view of moral and aesthetic values, it is also possible to err or to disagree. Just as in all forms of knowledge, aesthetic and moral knowledge is fallible. In addition, there is always room for reasonable disagreement on ethical issues. By extension, in aesthetics, there may be a whole range of 'aesthetically acceptable' solutions that leave room for subjective preferences. This is why it is impossible to predict aesthetic ideals that will be shared universally across time and space with certainty. But just as there can be an 'overlapping consensus' in ethics (Rawls, 1996), so can there be one in aesthetics. For example, various contributions in Nasar (1988) testify consensus in judgments about environmental aesthetics. Some moral and aesthetic options are off limits. In ethics, for example, the objective boundaries are clear violations of human rights (Wellman, 1963; Rachels, 1999; Moser and Carson,

2001; Roeser, 2006). In aesthetics, there are also clear cases where many people agree. Nasar (1988) provides empirical evidence for this based on a comparative study between American and Japanese subjects and their aesthetic judgments of urban street scenes. As Isaacs formulates it: "Each of us possesses an aesthetic instinct that is, at the most basic level, common to all, but molded by individual and cultural experience" (Isaacs, 2000, p. 154).

Theorists of architecture and aesthetics have proposed various design principles that are intended to contribute to the aesthetic quality of a building (Kaplan and Kaplan, 1982; Berlyne, 1974). Examples of these principles or features are order balanced against variation and disruption (Gombrich, 1984); or the balance between prospect (a wide view) and refuge (protection; Appleton, 1975). Folkmann (2009) proposes the following two aspects of aesthetics in design: design as a structure of sensual appearance; and design as an act of communication of an idea.

One problem is the frequent gap in aesthetic assessment between experts and laypeople (Hershberger, 1988; Gifford et al., 2000). In the case of an artwork in a museum, this is not so problematic as one can just avoid looking at the work in question. However, in the case of architecture or other aesthetic objects that are part of our lived environment, people are confronted with these objects every day. Hence, designers of such objects have the responsibility to take the aesthetic preferences of the people who occupy these public spaces into account. This includes the users of a building as much as viewers who regularly or incidentally pass by that building, as they are all more or less directly affected by its aesthetic aspects. This is the ethical dimension of aesthetics that I mentioned earlier. For this reason, architects cannot afford the same elitist aesthetic ideals as artists who create objects that people can avoid looking at (cf. Lampugnani, 2006).[6] Architecture is not '*l'art pour l'art*', but is an unavoidable part of our daily life and in that sense more closely related to consumer design. Fortunately, there is research into the aesthetic perceptions of laypeople that can help architects to consider the effect their designs have on laypeople (Hershberger and Cass, 1988; Groat, 1988; Gifford et al., 2000).

This understanding of aesthetics supports my idea that aesthetics fits well within risk discourse: there is uncertainty about universal aesthetic values but not complete ignorance. It might be impossible to quantify or assign probabilities to aesthetic values (although cf. Carlson, 1977). However, this holds for the other qualitative aspects of risk as well, and that is one of the aspects in which they differ from descriptive aspects of risk which can be measured and statistically monitored.

There are examples of architecture and urban planning that seem to succeed in being aesthetically sustainable. Of course, there is always room for disagreement and exceptions, as on all matters, and perhaps especially so on evaluative matters. However, there are examples of rather uncontroversial architecture such as the old downtown areas of many European cities, the temples in Tibet and other Asian countries, and the magnificent architecture of the Mayans and the Egyptian pyramids. These places attract millions of tourists year after year. Examples of modern buildings that incorporate a timeless aesthetic ideal include the Netherlands' 'Amsterdam school' buildings from the early 20th century and family homes from the 1930s that are still immensely popular (Kingma, 2016).

So, it does seem possible to build in an 'aesthetically sustainable way'. At the same time, it is a challenge for architects to remain aesthetically innovative. A safe, but not innovative way of building, is to imitate a well-established, popular style. In new suburbs in the Netherlands, there are currently a lot of retro-1930s houses or areas that imitate the famous Amsterdam canal houses of the 17th century. This seems to be 'the easy way out'. Here is an analogy with the precautionary principle from mainstream risk discourse: retro-architecture chooses to be 'safe rather than sorry'. But a well-known potential pitfall of the precautionary principle is that it might preserve a status quo that could actually be improved by new technological developments. As Sunstein (2005) has pointed out, holding on to the status quo can be just as risky as innovating. This insight from mainstream risk discourse can be directly applied to aesthetic risks: avoiding aesthetic innovation and its concomitant risks can lead to aesthetic conservatism, which is also a risk, by foreclosing the possibility of better designs. This means that avoiding one aesthetic risk can lead to another aesthetic risk. In the case of aesthetic aspects of architecture, architects can enhance people's experience by exploring new boundaries.

Similar ideas are discussed by van de Poel (2012) in the wider context of design for well-being. He distinguishes two main approaches to well-being and their relevance for engineering design. These approaches are desire satisfaction theories and objective list accounts. Van de Poel rejects the first account, on which designers should build what people desire, because people may desire things that are contrary to their own or other people's well-being (this resembles what I called populist approaches in chapter 2). Instead, he defends an objective list account in which designers have a moral responsibility to design for morally defensible accounts of well-being. These ideas can be extended to architecture: retro-architecture might satisfy the desires of people, but it does not provide them with anything that might in the end be even better (for example, d'Anjou, 2010, pleads for an authentic design ethics). Architecture can be innovative and yet pleasing to a large audience at the same time. Buildings can be designed to challenge our imagination and extend our ideas. In this way, architecture can diminish aesthetic risks and be 'aesthetically sustainable', without being attached to a status quo that would inhibit one of the features that makes architecture important for people: letting people explore new boundaries on what the buildings they live in and use could look like.

An important aspect of the precautionary principle is a reversal of the burden of proof. Rather than opponents having to prove that a new technology is dangerous, proponents have to show that it is safe. In the case of the aesthetic risks of architecture, architects should arguably do a good job to convince the public that their innovative designs are an improvement to more conventional designs that have proven their merit. This would include an educational task for architects—that is, to help people see the merits of new developments in architecture. Architects should show the public that their designs are aesthetically sustainable. They can be aided in this by architecture critics who might be better placed to articulate the merits of a building than the architects themselves.[7] If architects fail to convince the public, this might be a good reason to come up with something better.

This approach might raise concern that it could lead to paternalistic attitudes among architects. However, a normative approach to values in design can still incorporate the considerations of stakeholders without taking these to be ultimately authoritative. It could, for example, include an explicit phase of reflection and deliberation in the design process that takes into account stakeholders' values and allows for normative reflection (Manders-Huits and Zimmer, 2009). Emotions can play a constructive role in the exchange between architects and the public. Chapman (2009) develops the notion of 'emotional durability' for design that leads to less wasteful behavior by ensuring more stable relationships between consumers and products. Emotions can also play a role in designing for *aesthetic* sustainability. Empirical research indicates that it is hard for architects to predict laypeople's evaluations of buildings (Brown and Gifford, 2001). However, moral emotions can help architects to empathize and sympathize with potential users and to take into account their points of reference and experiences (cf. Greenbie, 1988, and Hershberger, 1988). This requires architects to train their emotions and imaginations so that they can put themselves into the shoes of a broad audience now and in the future. *Moral* emotions can help architects to keep in mind the well-being of the users of their buildings and of the people who will experience these buildings as part of their daily environment.

In addition, architects should use *aesthetic* emotions. *Aesthetic* emotions can help architects to create sublime designs; designs that can bring people into a state of aesthetic delight. Several authors who write on environmental aesthetics in general and on the aesthetics of architecture specifically emphasize the importance of emotions in the aesthetic assessment of architecture (Gifford et al., 2000, and several contributors to Nasar, 1988, make this point). In order to avoid the pitfall that architects get carried away by their own aesthetic emotions, there are methodologies which they can use to predict the aesthetic emotions of users. Desmet et al. (2007) have developed a methodology to measure and predict emotions elicited by industrial design products. This methodology could be extended to architecture.

There is an important role for policy makers here: they should design the institutional environment to allow for a dialogue between architects and the public. They could do this in several ways, such as organizing town hall meetings and providing information on new building projects in an accessible way. Policy makers have a special responsibility to enable a procedure that avoids the two pitfalls that threaten genuine deliberation and dialogue about new technologies that I discussed in previous chapters: that is, the technocratic pitfall where experts are invoked as the ones who provide the ultimate answers; and the populist pitfall where the will of the public is the ultimate arbiter. As I argued previously, neither of these situations comprises a genuine dialogue and exchange of ideas. Participatory approaches also usually insufficiently address emotions, and hence possible moral concerns. In Chapter 7, I developed an 'emotional deliberation approach to risk', which takes the emotions of stakeholders as a starting point for a genuine, critical deliberation on moral aspects of risk. In this approach, the outcomes are open, and a shift of positions is possible. Because all parties know that their concerns have received a fair hearing, they may be more willing to reach a consensus that may involve compromises from all parties. Aesthetic and moral emotions can help architects and policy makers to be sensitive to the aesthetic and moral concerns of people. By giving emotions an important role, 'aesthetic risks' can be minimized and open the way for aesthetically sustainable architecture.

8.6 How 'Techno-Art' Can Contribute to Emotional-Moral Deliberation

In the previous section, I discussed the importance of moral and aesthetic emotions and concerns in the context of architecture and aesthetic risks. However, there is another interesting contribution that aesthetics can make to emotional-moral reflection on risks.

The emotional deliberation approach to risk that I have developed in this book conceives of emotions as a *starting point* for moral discussion and reflection about risk. I have argued that emotions are not infallible but are nevertheless of crucial importance to moral reflection on technological risks. I have also argued that emotions can themselves be the subject and object of critical reflection; moral emotions can help us to critically reflect on more selfish emotions (Chapter 6). However, it can be difficult to transcend one's emotional-moral perspective. Emotions and moral views are shaped by the environment and culture in which people are raised. Emotions and moral views can resist influences that challenge people's core values (Kahan, 2012; Greene, 2013; Haidt, 2012). This can make public deliberation difficult. However, philosophers have argued that art can contribute to emotional-moral reflection (Levinson, 1998; Carroll, 2001; Nussbaum, 2001; Gaut, 2007; Bermúdez et al., 2006) and to politics (Adorno et al., 1980; Rorty, 1989; Groys, 2008; Bleiker, 2009; Kingston, 2011, p. 209; Kompridis, 2014). Art can provide meaning to our experiences via emotions (Slovic and Slovic, 2015). Art can help us transcend our given emotional-moral perspective by appealing to our imagination and compassion. For example, works of fiction can provide for sympathy, compassion and understanding for people whom one might initially experience as 'other' and in that sense frightening or disgusting. A novel such as *The Kite Runner* can generate sympathy and care for refugees, and the movie *Philadelphia* led to more acceptance and respect towards LGTBQ people in the United States. The question arises whether art can also make a positive contribution to emotional deliberation about technological risks.

There are numerous artists who are exploring the possibilities as well as the controversial aspects of new technologies in their work. This is what I call 'techno-art': art that reflects on and engages with risky technologies. However, these artworks have not yet been studied by philosophers. The aforementioned philosophers who study the role of art for emotional-moral and political reflection do not focus on techno-art but focus primarily on the context of interpersonal relationships and the development of personal virtues. Zwijnenberg (2009, 2014) is one of the few philosophers who studies bio-art, but he does not focus on other forms of techno-art, nor on risk and emotions. Scholars from the social sciences, literature, cultural and media studies who do study techno-art (Ede, 2000; Wilson, 2002; Pinsky, 2003; Anker and Nelkin, 2004; Da Costa and Philip, 2008; Weichman, 2008; Reichle, 2009; Munster, 2013) have not as yet focused specifically on the interrelationship between techno-art, moral emotions and risk. While there are empirical studies on the role of journalistic images, narratives and emotions in climate change risk perception (Leiserowitz, 2006; Spence and Pidgeon, 2010), and studies of the role of narrated scenarios for moral reflection on biotechnology (Boenink et al., 2010), these studies do not focus on art.

Hence, there is an important research gap, as techno-art can potentially make a constructive contribution to emotional-moral reflection on technological risks. In the remainder of this section, I will explore a number of issues related to this idea that deserve further research. But first I sketch a short historical outline of precursors of techno-art.

While many artists and writers have not even touched upon the subjects of science and technology, there are historical examples of artists who have engaged with scientific and technological developments. The most famous is probably Leonardo da Vinci, who made contributions to art as well as to science and technology. Others have taken a critical stance. Romanticism, for example, idealized a pre-industrialized world while dystopian novels examined the possible risks of technological developments. The latter still speak to the imagination. Aldous Huxley's *Brave New World* and Mary Shelley's *Frankenstein* articulated and shaped people's suspicions toward cloning and human enhancement. Furthermore, there have also been utopian artistic movements that explicitly embraced technology. These include several modernist movements at the beginning of the 20th century that celebrated technology and the efficiency and rationality it encompasses. Examples include De Stijl, Constructivism and Futurism, which were inspired by these developments. Science fiction is a well-established genre in literature and film dealing with utopian as well as dystopian visions of technological developments and their possible impacts on society. They probe ethical reflection through aesthetics and through the imaginative and emotional capacities of the audience. Science fiction author Isaac Asimov developed ethical guidelines—or so-called laws of robotics—implicitly and explicitly in his science fiction novels (cf. Asimov, 1950).

There are currently numerous artists and writers engaging with new technologies. The prominent writers Kazuo Ishiguro and Michel Houellebecq have written novels that explore a world in which cloning is a common established practice. There are new literary genres devoted to environmental as well as to climate issues, and there are even specialized academic journals devoted to the study of these new genres.

An increasing number of visual artists are also engaging with new technologies. These new technologies are providing for new material possibilities as well as opening a whole new array of topics to engage with. Examples are nuclear artist William Verstraeten; climate artists David Buckland and Boo Chapple; artists engaging with ICT and robotics, such as Stelarc; and bio artists Eduardo Kac, Anna Dumitriu, Jalila Essaidi and Patricia Piccinini.

In 2007, working at the intersection of bio art, robotic art and ICT art, performance artist Stelarc experimented with his own body by attaching a third ear to his arm by surgery and cell-cultivation, partially using his own stem cells. Eventually, there will be a microphone and a wireless connection to the Internet implanted in the ear. On his website, Stelarc explains the idea as follows.

> For example, someone in Venice could listen to what my ear is hearing in Melbourne. . . . Another alternate functionality [. . .] is the idea of the ear as part of an extended and distributed Bluetooth system. [. . .] This additional and enabled EAR ON ARM effectively becomes an Internet organ for the body.[8]

Through this work, Stelarc explores the potential of stem cell research and enhancement in a way that goes beyond the ways in which contemporary scientists usually approach developments in this field. He does it in an imaginative, playful and provocative way, exploring the technological and scientific possibilities and their legal and ethical boundaries. It is not directly clear what the use of EAR ON ARM would be, which is why scientists and technology developers would probably not even think about developing technologies in this way. Rather, it explores conceptual and normative issues, for example the meaning of connectedness in an age of hyperconnectivity.

The Young Family by Patricia Piccinini is a sculpture of a human-animal hybrid. It shows a mother and her infants in their bare flesh who look part human, part pig, part dog, and are described by Goriss-Hunter (2004) as "monstrous cute". While looking strange and repulsive, Donna Haraway writes about the care that Piccinini's creatures evoke. She sees them as symbolic of our modern technoculture in which we are obliged to care for what has been artificially created:

> Natural or not, good or not, safe or not, the critters of technoculture make a body-and-soul changing claim on their 'creators' that is rooted in the generational *obligation of* and *capacity for* responsive attentiveness. . . . To care is wet, emotional, messy, and demanding of the best thinking one has ever done. That is one reason we need speculative fabulation.
>
> (Haraway, 2011)

Haraway here refers to the emotions that are invoked by the artwork, but she also emphasizes the need for critical thinking emanating from this emotional experience.

The examples discussed above indicate that techno-artworks can reach out to people in a very direct way, by drawing on their imaginative and emotional capacities. Works of techno-art provide for concrete, vivid images and narratives that make technological developments and their possible impact on society more tangible for a larger audience than is possible via highly specialized scientific articles and abstract philosophical theorizing (Cf. Zwijnenberg, 2009, p. 19). Techno artists can create awareness of societal issues (Gessert, 2003) and provide for critical reflection on technological and scientific developments (Reichle, 2009; Zwijnenberg, 2009). This is an area that is still largely unexplored by philosophers and scholars who study societal aspects of risky technologies.

Philosophers have developed accounts on how art in general, not specifically techno-art, helps our ethical and political thinking (e.g., Carroll, 2001; Nussbaum, 2001; Gaut, 2007; Adorno et al., 1980; Rorty, 1989; Kompridis, 2014). These philosophical approaches have mainly focused on more traditional works of art that engage with interpersonal and societal relationships. However, works of techno-art are materially and in terms of content different from traditional works of art, giving rise to different philosophical questions. Questions that techno-art might raise include the aspects of a technological risk that the artwork addresses; the moral values of the technology exemplified or used in the artwork; and how the artwork addresses or highlights these values.

Further, an important aspect that is not yet discussed in detail in the limited academic literature on techno-art is that these artworks often give rise to emotional responses. The question then emerges as to the role that emotions might play in the ethical reflection triggered by the works of techno-art. Stelarc's work might give rise to fascination about the possibilities of technology, as well as to disgust at the gross image, or annoyance at the seemingly decadent endeavor. What kind of values and concerns do these emotions point to, and how should we assess them? The work by Piccinini gives rise to feelings of care as well as to feelings of discomfort and uncanniness. The latter can point to the unclear moral status of artificial life and human-animal hybrids and our undefined moral responsibilities toward them. Of course, much more can, and needs to, be said about these works of art. These examples illustrate that works of techno-art provide rich material that requires hermeneutical, in-depth studies of artworks as well as normative-ethical reflection on the emotions, and aesthetic and reflective experiences that these works give rise to. These can ultimately provide us with deeper insights into ethical aspects of new technologies.

There are numerous approaches to participatory technology assessment that aim at providing for more democratic decision-making on risky technologies (Cf. van Asselt and Rijkens-Klomp, 2002, cf. Chapter 2). Works of techno-art can provide us with direct experiences with new technologies and help us probe our emotional-moral reflection. Hence, they can play a powerful role in public deliberation. This idea has not yet received much attention either in the academic literature or in practical methodologies for public deliberation on risky technologies.

For example, works of art can play a crucial role in the debate on sustainable energy technologies and on combatting climate change. They can help by making climate change more salient, inciting people to take action (see section 8.3), and letting people critically reflect on the possible role of climate engineering. Artist Boo Chappel asked people to wear reflective white hats, symbolizing ideas from geo-engineering to protect the earth with a reflective layer to prevent global warming. Chappel then asked people to report and reflect on their very tangible, symbolic experiences and on the impact that this technology would have. The works of Stelarc and Piccinini exemplify the ambiguous feeling that many people have of developments in biotechnology in a direct, experiential way. A common concern is that biotechnologists 'play with nature', but this is a complex notion that is difficult to conceptualize in a purely analytical way. Artists can play with the public's uncertainties and uneasy feelings by developing artworks via biotechnology and by examining the boundaries between life and technology. Such artworks may not necessarily provide for clear-cut answers, but they can provide for a common platform for reflection where different kinds of stakeholders can deliberate together on an equal footing. Artworks have the capacity to make complex scientific developments concrete and tangible for a larger audience.

Another area where techno-art can make an important contribution is in critical societal reflection on robotics, AI and ICT. Information technologies are deeply ingrained in our contemporary societies and bring many conveniences. This is the reason that many people endorse specific ICT technologies, such as computers or smartphones. However, ICTs can also lead to massive privacy intrusions. Increased automation may change our labor markets for good by making large parts of society obsolete on the work floor. There are concerns about artificial intelligences getting out of control and eventually taking over from humans. That this is not confined to vastly exaggerated science fiction novels and movies is exemplified by the fact that numerous leading scholars and technologists, including Stephen Hawking and Elon Musk, recently signed a letter warning against the dangers of AI and demanding research on how to achieve 'beneficial artificial intelligence' (Future of Life Institute, 2015). Artists who work with artificial intelligence and robotics can play an important role in critically reflecting on what beneficial artificial intelligence could be. They can explore the possibilities of beneficial artificial intelligence in more accessible, real-life settings than scientific laboratories before any artificial intelligence and robotics are introduced into society.

In this way, techno-art can contribute to overcoming the so-called Collingridge dilemma (Collingridge, 1980), which states that the potentially detrimental effects of technologies will only be fully understood when these technologies are used by society, by which time regulation might be too late and the technologies may already have had negative side effects. On the other hand, acting preventively by restricting technology by regulation before it is part of society might lead to wrong estimates and the withholding of society of potentially useful applications. Ibo van de Poel (2013) has proposed viewing technology as a social experiment with explicit rules for aborting the process should well-grounded ethical concerns arise. However, even then undesirable, irreversible effects might already have taken place. Techno-art provides for an additional route before technologies are introduced into society on a large scale. Techno-artists can go beyond the limitations of the labs of scientists and technology developers. They can take technologies to greater extremes and explore different scenarios in more tangible ways, and their works can be more accessible to society.

Next to these promises of techno-art, there are also possible concerns. Techno-art can potentially make an important contribution to public debates, by highlighting societal implications, emotions and values related to new technologies. Artworks can inspire emotional-moral reflection on risky technologies. However, techno-art can also fail to do these things. The artworks might be hard to grasp, appear manipulative or focus on unrealistic scenarios. Nevertheless, even in these cases artworks might function as a trigger for emotional reflection, deliberation and discussion.

Another concern is that while a work of techno-art might be helpful in emotional-ethical reflection, it might not be clear what its artistic and aesthetic merits are. For example, some people debate whether Stelarc's project should be considered art and, if so, in what respect?

Furthermore, there is an area of possible tension between artistic freedom on the one hand and explicitly inviting artists to play a role in public deliberation on risky technologies on the other. It can be difficult to ensure that artists have the freedom to pursue their own ideas while at the same time making meaningful contributions to ethical reflection and doing justice to scientific and legal constraints.

This relates to another question: should artists be bound by the same ethical and legal constraints as scientists and technology developers? As artists, they might arguably deserve more freedom, also given that their works are unlikely to be produced on a large scale. Alternatively, one might argue that while technologies might contribute directly to societal well-being, this might be less evident in the case of artworks. This may make artworks less useful, and consequently, they may not be given the same space to introduce possible risks. Yet another position could be that artists and scientists deserve the same space for exploration without endangering the public, and should thus be subject to the same ethical and legal restrictions. There are as yet no guidelines for these issues, but given the fact that more and more artists are involved with risky technologies, these are fascinating questions that require further research (Roeser et al., forthcoming).

To conclude, I think that techno-art can potentially make a constructive contribution to emotional-moral reflection on the risks of technological developments. Techno-art can help people to make abstract problems concrete; explore new scenarios; challenge their imaginations; and broaden their narrow personal perspectives through empathy, sympathy and compassion. Techno-art can contribute to an emotional-moral reflection about the kind of society we might want to live in. This means that techno-art can potentially contribute to public debate and overcoming stalemates. At the same time, techno-art faces various challenges such as how to preserve the non-instrumental nature of art while playing a role in public debates, and how to do so in a meaningful way. Techno-art is a rapidly expanding field of artistic development that could potentially make a major contribution to emotional deliberation, responsible innovation and therefore society, but this requires thorough investigation by philosophers and other scholars.

8.7 Conclusion

In this chapter I have discussed several ways of incorporating emotions in political decision making about risky technologies, based on the new approach to risk emotions developed in the previous chapters of this book. This approach allows for a different way of dealing with risk emotions in public debates by avoiding both technocratic and populist pitfalls alike. Instead, this alternative approach allows for what I call an 'emotional deliberation approach to risk.' It allows the public a genuine voice in which their emotions and concerns are appreciated, listened to and discussed, rather than ignored (technocratic pitfall) or taken as a given that makes discussion impossible (populist pitfall). This approach builds on and expands existing approaches to participatory risk assessment by explicitly including emotions.

In this chapter, I have examined the possibilities of this approach by discussing a variety of technological risks, related to nuclear energy, climate change, health technologies and architecture. I have argued that discussing emotions and their underlying concerns contributes to a well-grounded ethical assessment of risky technologies. At the same time, this approach can help to overcome the gap between experts and laypeople that occurs repeatedly in debates about risky technologies. It can provide for circumstances in which all involved parties respect each other more and are willing to listen to each other and compromise in a debate. This is necessary in order to deliberate on morally responsible and meaningful ways of dealing with risky technologies. Furthermore, risk emotions can provide for motivation to change one's behavior toward a more sustainable lifestyle, as in the case of climate change. Emotions can also provide insights into aesthetic aspects of technologies, as in the case of architecture and urban planning. In addition, aesthetics can provide us with further ways to emotionally reflect on risky technologies, namely in the case of art that engages with risky technologies. Hence, incorporating risk emotions promises to overcome several problems that currently haunt debates about risky technologies. Risk emotions enable us to see moral values and to act accordingly; and they can help to overcome stalemates.

Notes

1. Nevertheless, Zagzebski acknowledges the possibility that people have motivational responses to abstract moral principles without necessarily being emotional about them, cf. Zagzebski (2003, p. 122).
2. In Chapter 5 of my book Roeser (2011a), I discuss this topic at greater length.
3. www.gezondheidsnet.nl/baarmoederhalskanker/groot-aantal-meisjes-haalt-geen-hpv-vaccinatie
4. For environmental aesthetics and the role of architecture, cf. Nasar (1988) and Carlson (2000); on aesthetics in architecture cf. Scruton (1979) and Hill (1999).
5. Exceptions of studies that mention aesthetics as a determinant in risk perception are Willis and Dekay (2005) and Willis et al. (2007).
6. Thanks to Lara Schrijver for drawing my attention to this publication.
7. Thanks to Lara Schrijver for suggesting this point to me.
8. See http://stelarc.org/?catID=20242.

Epilogue

This book has explored the role emotions do and can play in debates about risky technologies. Most authors who write on risk emotions see them as a threat to rational decision making about risks. However, based on recent developments in emotion research, an alternative picture of risk emotions is possible. I argue that risk emotions can be an important source of insight into morally salient aspects of risk. However, this does not make moral judgments concerning risks subjective, as has been claimed by psychologists and social scientists. Rather, moral emotions are an indispensable source of awareness of ethical aspects of risk. This view allows for fruitful insights on how to improve public debates about risk, and to overcome the gap between experts and laypeople that currently so often leads to a deadlock in discussions about risky technologies. A new understanding of risk emotions as source of practical rationality can make a positive contribution to various theoretical and practical conundrums. It can solve what I call the 'puzzle of lay rationality', and it can provide for a more fruitful approach to public decision making about risks than current alternatives. Policy making about risky technologies should do justice to emotions as an invaluable source of ethical insight. Emotions should not be neglected or seen as a "given" that cannot be investigated any further, but they should be a trigger for discussion. In democratic decision-making procedures about risk, the emotions of the public should be taken seriously. Purely rational decision procedures overlook important ethical considerations about risks. Instead, my argument is that the emotions of the public should be taken seriously in order to arrive at well-grounded judgments about the moral acceptability of risks. I have discussed a variety of risky technologies to illustrate the potential of this approach. As I said at the beginning of this book, my aim is to be as charitable as possible to our human capacities. I hope to have been able to provide some perspective on how the maybe most controversial of our capacities, namely our intuitions and emotions, can actually provide us with important insights in the highly specialized area of risky technologies. Our moral emotions and intuitions are an important moral compass and source of reflection and understanding that we should involve in critical public deliberation about risky technologies, a subject matter that is too important to miss out on crucial ways to make more well-grounded and more democratic decisions.

References

Adorno, T., Benjamin, W., Bloch, E., Brecht, B., and Lukacs, G. (1980). *Aesthetics and Politics*. New York: Verso.

Ahteensuu, M. and Sandin, P. (2012). The precautionary principle. In S. Roeser, R. Hillerbrand, P. Sandin, and M. Peterson (eds.), *Handbook of Risk Theory*. Dordrecht: Springer, pp. 961–978.

Alfano, M., Huijts, N., and Roeser, S. (forthcoming). Trust in politics, institutions and governance. In Judith Simon (ed.), *Routledge Handbook of Trust and Philosophy*. London: Routledge.

Alhakami, A.S. and Slovic, P. (1994). A psychological study of the inverse relationship between perceived risk and perceived benefit. *Risk Analysis*, 14(6): 1085–1096.

Alston, W.P. (1989). *Epistemic Justification: Essays in the Theory of Knowledge*. Ithaca and London: Cornell University Press.

Alston, W.P. (1993). *The Reliability of Sense Perception*. Ithaca and London: Cornell University Press.

Anker, S. and Nelkin, D. (2004). *The Molecular Gaze. Art in the Genetic Age*. New York: Cold Spring Harbor Laboratory Press.

Appleton, J. (1975). *The Experience of Landscape*. London: William Clowes.

Asimov, I. (1950). *I, Robot*. New York: Gnome Press.

Asveld, L. (2007). Autonomy and risk: Criteria for international trade regimes. *Journal of Global Ethics*, 3(1): 21–38.

Asveld, L. (2008). *Respect for Autonomy and Technological Risks*. PhD-thesis, TU Delft.

Asveld, L. (2009). Trust and criteria for proof of risk: The case of mobile phone technology in The Netherlands. In L. Asveld and S. Roeser (eds.), *The Ethics of Technological Risk*. London: Earthscan, pp. 220–234.

Asveld, L. and Roeser, S. (eds.) (2009). *The Ethics of Technological Risk*. London: Earthscan.

Audi, R. (2003). *The Good in the Right: A Theory of Intuition and Intrinsic Value*. Princeton: Princeton University Press.

Audi, R. (2013). *Moral Perception*. Princeton: Princeton University Press.

Basta, C. (2012). Risk and Spatial Planning. In S. Roeser, R. Hillerbrand, M. Peterson and P. Sandin (eds.), *Handbook of Risk Theory: Epistemology, Decision Theory, Ethics, and Social Implications of Risk*. Dordrecht: Springer, pp. 265–294.

Beauchamp, T. and Childress, J. (2006). *Principles of Biomedical Ethics*. Oxford: Oxford University Press.

Bedke, M.S. (2008). Ethical intuitions: What they are, what they are not, and how they justify. *American Philosophical Quarterly*, 43(3): 253–270.

Bentham, J. (2007 [1780]). *An Introduction to the Principles of Morals and Legislation*. Mineola: Dover Publications.
Ben-Ze'ev, A. (2000). *The Subtlety of Emotions*. Cambridge: MIT Press.
Berlyne, D.E. (1974). *Studies in the New Experimental Aesthetics: Steps Toward an Objective Psychology of Aesthetic Appreciation*. New York: Halsted.
Bermúdez, J.L. and Gardner, S. (eds.) (2006). *Art and Morality*. London: Routledge.
Blackburn, S. (1998). *Ruling Passions*. Oxford: Oxford University Press.
Bleiker, R. (2009). *Aesthetics and World Politics*. Basingstoke: Palgrave Macmillan.
Blum, L.A. (1994). *Moral Perception and Particularity*. Cambridge and New York: Cambridge University Press.
Boenink, M., Swierstra, T., and Stemerding, D. (2010). Anticipating the interaction between technology and morality: A scenario study of experimenting with humans in bionanotechnology. *Studies in Ethics, Law, and Technology*, 4(2): 4. Available at: https://doi.org/10.2202/1941-6008.1098
Bohman, J. and Rehg, W. (eds.) (1997). *Essays on Reason and Politics: Deliberative Democracy*. Boston: MIT Press.
Bohnenblust, H. and Slovic, P. (1998). Integrating technical analysis and public values in risk-based decision making, *Reliability Engineering & System Safety*, 59: 151–159.
Brady, M. (2013). *Emotional Insight: The Epistemic Role of Emotional Experience*. Oxford: Oxford University Press.
Broad, C.D. (1951 [1930]). *Five Types of Ethical Theory*. London: Routledge and Kegan Paul.
Brown, G. and Gifford, R. (2001). Architects predict lay evaluations of large contemporary buildings: Whose conceptual properties? *Journal of Environmental Psychology*, 21: 93–99.
Buck, R. and Davis, W. (2010). Marketing risks: The mindless acceptance of risks is promoted by emotional appeals. In S. Roeser (ed.), *Emotions and Risky Technologies*. Dordrecht: Springer, pp. 61–80.
Buck, R. and Ferrer, R. (2012). Emotion, warnings, and the ethics of risk communication. In S. Roeser, R. Hillerbrand, M. Peterson and P. Sandin (eds.), *Handbook of Risk Theory: Epistemology, Decision Theory, Ethics, and Social Implications of Risk*. Dordrecht: Springer, pp. 693–723.
Carlson, A. (1977). On the possibility of quantifying scenic beauty. *Landscape Planning*, 4: 131–172.
Carlson, A. (2000). *Aesthetics and the Environment: The Appreciation of Nature, Art and Architecture*. London: Routledge.
Carlson, A. (2007). Environmental aesthetics. In E.N. Zalta (ed.), *The Stanford Encyclopedia of Philosophy*. SEP, Stanford. Available at: http://plato.stanford.edu/entries/environmental-aesthetics/
Carroll, N. (2001). *Beyond Aesthetics: Philosophical Essays*. Cambridge: Cambridge University Press.
Chapman, J. (2009). Design for (emotional) durability. *Des Issues*, 25: 29–35.
Cohen, B.L. (1998). Public perception versus results of scientific risk analysis. *Reliability Engineering & System Safety*, 59: 101–105.
Collingridge, D. (1980). *The Social Control of Technology*. New York: St. Martin's Press and London: Pinter.
Correljé, A., Cuppen, E., Dignum, M., Pesch, U., and Taebi, B. (2015). Responsible Innovation in Energy Projects: Values in the Design of Technologies, Institutions

and Stakeholder Interactions. In B.J. Koops, I. Oosterlaken, H. Romijn, T. Swierstra and J. van den Hoven (eds.), *Responsible Innovation 2*. Dordrecht: Springer, pp. 183–200.

Cross, F.B. (1998). Facts and values in risk assessment. *Reliability Engineering & System Safety*, 59: 27–40.

Cullison, A. (2010). What are Seemings? *Ratio*, 23: 260–274.

Cuneo, T. (2007). *The Normative Web: An Argument for Moral Realism*. Oxford: Oxford University Press.

Cuppen, E. (2012). Diversity and constructive conflict in stakeholder dialogue: Considerations for design and methods. *Policy Sciences*, 45(1): 23–46.

Cuppen, E., Brunsting, S., Pesch, U., and Feenstra, Y. (2015). How stakeholder interactions can reduce space for moral considerations in decision making: A contested CCS project in the Netherlands. *Environment and Planning A*, 47: 1963–1978.

Da Costa, B. and Philip, K. (eds.) (2008). *Tactical Biopolitics: Art, Activism, and Technoscience*. Cambridge: The MIT Press.

Damasio, A.R. (1994). *Descartes' Error: Emotion, Reason and the Human Brain*. New York: G.P. Putnam.

Dancy, J. (1993). *Moral Reasons*. Oxford: Blackwell.

Dancy, J. (2004). *Ethics Without Principles*. Oxford and New York: Clarendon Press and Oxford University Press.

Dancy, J. (2014). Intuition and emotion. *Ethics*, 124(4): 787–812.

d'Anjou, P. (2010). Beyond duty and virtue in design ethics. *Des Issues*, 26: 95–105.

De Hollander, A.E.M. and Hanemaaijer, A.H. (2003). *Nuchter omgaan met risico's: Milieu—en natuurplanbureau (MNP)—RIVM*. Bilthoven: RIVM.

Deonna, J.A. and Teroni, F. (2012). *The Emotions: A Philosophical Introduction*. New York: Routledge.

Desmet, P.M.A., Porcelijn, R., and van Dijk, M. (2007). Emotional design; Application of a research based design approach. *Journal of Knowledge Technology Policy*, 20(3): 141–155.

de Sousa, R. (1987). *The Rationality of Emotion*. Cambridge: MIT Press.

de Sousa, R. (2003). Emotion. In E.N. Zalta (ed.), *The Stanford Encyclopedia of Philosophy*. Available at: http://plato.stanford.edu/archives/spr2003/entries/emotion/

de Sousa, R. (2010). Here's how I feel: Don't trust feelings. In S. Roeser (ed.), *Emotions and Risky Technologies*. Dordrecht: Springer, pp. 17–35.

Dreyfus, H. (1992). *What Computers Still Can't Do: A Critique of Artificial Reason*. Cambridge: MIT Press.

Ede, S. (ed.) (2000). *Strange and Charmed: Science and the Contemporary Visual Arts*. Preface by A.S. Byatt. London: Calouste Gulbenkian Foundation.

Eisenberg, A.E., Baron, J., and Seligman, M.E.P. (1998). Individual differences in risk aversion and anxiety. *Psychological Bulletin*, 87: 245–251.

Ellsberg, D. (1961). Risk, Ambiguity, and the Savage Axioms. *The Quarterly Journal of Economics*, 75: 643–669.

Enoch, D. (2011). *Taking Morality Seriously: A Defense of Robust Realism*. Oxford: Oxford University Press.

Epstein, L.G. (1999). A definition of uncertainty aversion. *Review of Economic Studies*, 66(3): 579–608.

Epstein, S. (1994). Integration of the cognitive and the psychodynamic unconscious. *American Psychologist*, 49(8): 709–724.

Espinoza, N. (2009). Incommensurability: The failure to compare risks. In L. Asveld and S. Roeser (eds.), *The Ethics of Technological Risk*. London: Earthscan/Routledge, pp. 128–143.

Ewing, A.C. (1929). *The Morality of Punishment, With Some Suggestions for a General Theory of Ethics*. London: Kegan Paul and Trench: Trubner & Co., Ltd.

Ewing, A.C. (1941). Reason and intuition. *Proceedings of the British Academy*, 27: 67–107.

Faulkner, W. (2000). Dualisms, hierarchies and gender in engineering. *Social Studies of Science*, 30(5): 759–792.

Feenstra, C.F.J., Mikunda, T., and Brunsting, S. (2010). *What Happened in Barendrecht? Case Study on the Planned Onshore Carbon Dioxide Storage in Barendrecht, the Netherlands*. ECN Report.

Fetherstonhaugh, D., Slovic, P., Johnson, S., and Friedrich, J. (1997). Insensitivity to the value of human life: A study of psychophysical numbing. *Journal of Risk and Uncertainty*, 14: 283–300.

Finucane, M. (2012). The role of feelings in perceived risk. In S. Roeser, R. Hillerbrand, M. Peterson and P. Sandin (eds.), *Handbook of Risk Theory: Epistemology, Decision Theory, Ethics, and Social Implications of Risk*. Dordrecht: Springer, pp. 677–691.

Finucane, M., Alhakami, A., Solvic, P., and Johnson, S.M. (2000). The affect heuristic in judgments of risks and benefits. *Journal of Behavioral Decision Making*, 13: 1–17.

Fischhoff, B. (1996). Public values in risk research. Challenges in risk assessment and risk management. *Annals of the American Academy of Political and Social Science*, 545: 73–84.

Fischhoff, B., Lichtenstein, S., Slovic, P., Derby, S.L., and Keeney, R. (1981). *Acceptable Risk*. Cambridge: Cambridge University Press.

Fischhoff, B., Slovic, P., Lichtenstein, S., Read, S., and Combs, B. (1978). How safe is safe enough? A psychometric study of attitudes towards technological risks and benefits. *Policy Science*, 9: 127–152.

Folkmann, M.N. (2009). Evaluating aesthetics in design: A phenomenological approach. *Design Issues*, 26: 40–53.

Frank, R. (1988). *Passions Within Reason: The Strategic Role of the Emotions*. New York: W. W. Norton.

Friedman, B. (2004). Value sensitive design. In W.S. Bainbridge (ed.), *Encyclopedia of Human-Computer Interaction*. Great Barrington: Berkshire Publishing Group, pp. 769–774.

Frijda, N. (1986). *The Emotions*. Cambridge: Cambridge University Press.

Future of Life Institute. (2015). *Research Priorities for Robust and Beneficial Artificial Intelligence*. Available at: http://futureoflife.org/ai-open-letter/

Gaudine, A. and Thorne, L. (2001). Emotion and ethical decision-making in organizations. *Journal of Business Ethics*, 31: 175–187.

Gaut, B. (2007). *Art, Emotion and Ethics*. Oxford: Oxford University Press.

Gerrans, P. and Kennett, J. (2006). Introduction: Is cognitive penetrability the mark of the moral? *Philosophical Explorations* (Special Issue: Empirical Research and the Nature of Moral Judgment), 9(1): 3–12.

Gessert, G. (2003). Notes on the art of plant breeding. In J. Hauser (ed.), *L'Art Biotech Catalogue*, exhibition catalog. Nantes: Le Lieu Unique, p. 47.

Gettier, E. (1963). 'Is Justified True Belief Knowledge?', Analysis 23: 121–123.

Gibbard, A. (1990). *Wise Choices, Apt Feelings*. Cambridge: Harvard University Press.

Gifford, R., Hine, D.H., Muller-Clemm, W., Dárcy, J., Reynolds, J.R., and Shaw, K.T. (2000). Decoding modern architecture: A lens model approach for understanding the aesthetic differences of architects and laypersons. *Environment & Behavior*, 32: 163–186.

Gigerenzer, G. (2002). *Reckoning With Risk*. London: Penguin.

Gigerenzer, G. (2007). *Gut Feelings: The Intelligence of the Unconscious*. London: Viking.

Gilovich, T., Griffin, D., and Kahneman, D. (Eds.) (2002). *Heuristics and Biases: The Psychology of Intuitive Judgement*. New York: Cambridge University Press.

Goldie, P. (2000). *The Emotions: A Philosophical Exploration*. Oxford and New York: Clarendon Press.

Gombrich, E.H. (1984). *The Sense of Order*. London: Phaidon.

Goriss-Hunter, A. (2004). Slippery mutants perform and wink at maternal insurrections: Patricia Piccinini's monstrous cute. *Continuum: Journal of Media & Cultural Studies*, 18: 541–553.

Green, O.H. (1992). *The Emotions: A Philosophical Theory*. Dordrecht and Boston: Kluwer.

Greenbie, B.B. (1988). The landscape of social symbols. In J.L. Nasar (ed.), *Environmental Aesthetics: Theory, Research, and Applications*. Cambridge: Cambridge University Press, pp. 64–73.

Greene, J.D. (2003). From neural 'is' to moral 'ought': What are the moral implications of neuroscientific moral psychology? *Nature Reviews Neuroscience*, 4: 847–850.

Greene, J.D. (2007). The secret joke of Kant's soul. In W. Sinnott-Armstrong (ed.), *Moral Psychology. Volume 3: The Neuroscience of Morality: Emotion, Disease, and Development*. Cambridge: MIT Press, pp. 2–79.

Greene, J.D. (2013). *Moral Tribes*. New York: Penguin.

Greene, J.D. and Haidt, J. (2002). How (and where) does moral judgment work? *Trends in Cognitive Sciences*, 6: 517–523.

Greene, J.D., Nystrom, L.E., Engell, A.D., Darley, J.M., and Cohen, J.D. (2004). The neural bases of cognitive conflict and control in moral judgment. *Neuron*, 44: 389–400.

Greene, J.D., Sommerville, R.B., Nystrom, L.E. Darley, J.M., and Cohen, J.D. (2001). An fMRI investigation of emotional engagement in moral judgment. *Science*, 293: 2105–2108.

Greenspan, P.S. (1988). *Emotions and Reasons: An Inquiry into Emotional Justification*. New York and London: Routledge.

Gregory, R. and Keeney, R.L. (1994). Creating policy alternatives using stakeholder values. *Management Science*, 40(8): 1035–1048.

Griffith, P.E. (1997). *What Emotions Really Are: The Problem of Psychological Categories*. Chicago: University of Chicago Press.

Groat, L.N. (1988). Contextual compatibility in architecture: An issue of personal taste? In J.L. Nasar (ed.), *Environmental Aesthetics: Theory, Research, and Applications*. Cambridge: Cambridge University Press, pp. 228–253.

Groys, B. (2008). *Art Power*. Cambridge: MIT Press.

Gutmann, A. and Thompson, D. (2000). Why deliberative democracy is different. *Social Philosophy and Policy*, 17(1): 161–180.

Haber, G., Malow, R.M., and Zimet, G.D. (2007). The HPV Vaccine Mandate Controversy. *Journal of Pediatric and Adolescent Gynecology*, 20: 325–331.

Habermas, J. (1985). *The Theory of Communicative Action. Volume 2: Lifeword and System: A Critique of Functionalist Reason* (Vol. 2). Boston: Beacon Press.

Habermas, J. (1996). Three normative models of democracy. In S. Benhabib (ed.), *Democracy and Difference: Contesting the Boundaries of the Political*. Princeton: Princeton University Press, pp. 189–208.

Haidt, J. (2001). The emotional dog and its rational tail: A social intuitionist approach to moral judgment. *Psychological Review*, 108(4): 814–833.

Haidt, J. (2012). *The Righteous Mind: Why Good People Are Divided by Politics and Religion*. New York: Vintage Books.

Haidt, J. and Graham, J. (2007). When morality opposes justice: Conservatives have moral intuitions that liberals may not recognize. *Social Justice Research*, 20: 98–116.

Hall, C. (2005). *The Trouble With Passion: Political Theory Beyond the Reign of Reason*. New York: Routledge.

Hansson, S.O. (2003). Are natural risks less dangerous than technological risks? *Philosophia Naturalis*, 40: 43–54.

Hansson, S.O. (2004). Philosophical perspectives on risk. *Techné*, 8: 10–35.

Haraway, D. (2011). Speculative fabulations for technoculture's generations: Taking care of unexpected country. *Australian Humanities Review*, 50. Available at: http://www.australianhumanitiesreview.org/archive/Issue-May-2011/haraway.html.

Harries, K. (1997). *The Ethical Function of Architecture*. London: MIT Press.

Harvey, M. (2009). Drama, talk, and emotion: omitted aspects of public participation. *Science, Technology & Human Values*, 34(2): 139–161.

Hayenhjelm, M. (2012). What is a Fair Distribution of Risk? In S. Roeser, R. Hillerbrand, M. Peterson and P. Sandin (eds.), *Handbook of Risk Theory*. Dordrecht: Springer, pp. 909–929.

Held, V. (2006). *The Ethics of Care: Personal, Political, and Global* (2nd ed.). Oxford and New York: Oxford University Press.

Hendrickx, L., Vlek, C., and Oppeval, H. (1989). Relative importance of scenario information and frequency information in the judgment of risk. *Acta Pyschologica*, 72: 41–63.

Hermansson, H. (2011). Defending the conception of 'objective risk'. *Risk Analysis*, 32: 16–24.

Hershberger, R.G. (1988). A study of meaning and architecture. In J.L. Nasar (ed.), *Environmental Aesthetics: Theory, Research, and Applications*. Cambridge: Cambridge University Press, pp. 175–194.

Hershberger, R.G. and Cass, R.C. (1988). Predicting user responses to buildings. In J.L. Nasar (ed.), *Environmental Aesthetics: Theory, Research, and Applications*. Cambridge: Cambridge University Press, pp. 195–211.

Hill, R. (1999). *Designs and Their Consequences: Architecture and Aesthetics*. New Haven and London: Yale University Press.

Hoggett, P. and Thompson, S. (2002). Toward a democracy of the emotions. *Constellations*, 9(1): 106–126.

Huemer, M. (2005). *Ethical Intuitionism*. Basingstoke: Palgrave Macmillan.

Hulme, M. (2009). *Why We Disagree About Climate Change*. Cambridge: Cambridge University Press.

Hume, D. (1975). *A Treatise of Human Nature* (2nd ed., revised by P.H. Nidditch). Edited by L.A. Selby-Bigge. Oxford: Clarendon Press.

Isaacs, R. (2000). The urban picturesque: An aesthetic experience of urban pedestrian places. *Journal of Urban Design*, 5: 145–180, p. 154.

Isen, A.M. (1993). Positive affect and decision making. In M. Lewis and J. Haviland (eds.), *Handbook of Emotion*. New York: Guilford Press, pp. 261–277.

Jaeger, C.C., Webler, T., Eugene, A.R., and Renn, O. (2001). *Risk, Uncertainty, and Rational Action*. London: Earthscan.

Jasanoff, S. (1993). Bridging the two cultures of risk analysis. *Risk Analysis*, 13: 123–129.

Jasanoff, S. (1998). The political science of risk perception. *Reliability Engineering & System Safety*, 59(1): 91–99.

Johnson, E.J. and Amos, T. (1983). Affect, generalization, and the perception of risk. *Journal of Personality and Social Psychology*, 45: 20–31.

Kahan, D.M. (2000). The progressive appropriation of disgust. In S. Bandes (ed.), *The Passions of Law*. New York: NYU Press.

Kahan, D.M. (2008). Two conceptions of emotion in risk regulation. *University of Pennsylvania Law Review*, 156: 741–766.

Kahan, D.M. (2012). Cultural cognition as a conception of the cultural theory of risk. In S. Roeser, R. Hillerbrand, M. Peterson, and P. Sandin (eds.), *Handbook of Risk Theory*. Dordrecht: Springer, pp. 725–759.

Kahan, D.M. and Slovic, P. (2006). *Cultural Evaluations of Risk: 'Values' or 'Blunders'?* Yale Law School, Public Law Working Paper No. 111.

Kahan, D.M., Slovic, P., Braman, D., and Gastil, J. (2006). Fear of Democracy: A Cultural Evaluation of Sunstein on Risk. A review of *Laws of Fear: Beyond the Precautionary Principle* by Cass R. Sunstein. *Harvard Law Review*, 119: 1071–1109.

Kahane, G. and Shackel, N. (2010). Methodological issues in the neuroscience of moral judgement. *Mind and Language*, 25: 561–582.

Kahneman, D. (2011). *Thinking Fast and Slow*. New York: Farrar, Straus and Giroux.

Kaliarnta, S. (2016). Using Aristotle's theory of friendship to classify online friendships: a critical counterview. *Ethics and Information Technology*, 18: 65–79.

Kaliarnta, S., Fahlquist, J.N., and Roeser, S. (2011). Emotions and ethical considerations of women undergoing IVF-treatments. *Health Care Ethics Committees Forum*, 23(4): 281–293.

Kant, I. (1964 [1786]). *Groundwork of the Metaphysic of Morals*. New York: Harper and Row.

Kaplan, S. and Kaplan, R. (1982). *Cognition and the Environment: Functioning in an Uncertain World*. New York: Praeger.

Kass, L. (1997). The wisdom of repugnance: Why we should ban the cloning of human beings. *The New Republic* (June 2, 1997).

Kingma, J. (2016). *De magie van het jaren 30 huis*. Nijmegen: Vantilt.

Kingston, R. (2011). *Public Passion: Rethinking the Grounds for Political Justice*. Montreal: McGill-Queen's University Press.

Kingston, R. and Ferry, L. (eds.) (2008). *Bringing the Passions Back In: The Emotions in Political Philosophy*. British Columbia: University of British Columbia Press.

Klein, S. (2002). The head, the heart, and business virtues. *Journal of Business Ethics*, 39: 347–359.

Koenigs, M., Kruepke, M., Zeier, J., and Newman, J.P. (2012). Utilitarian moral judgment in psychopathy. *Social Cognitive and Affective Neuroscience*, 7: 708–714.

Kompridis, N. (ed.) (2014). *The Aesthetic Turn in Political Thought*. London: Bloomsbury Academic.

Koons, J.R. (2003). Why response-dependence theories of morality are false. *Ethical Theory and Moral Practice*, 6: 275–294.

Korsgaard, C. (1996a). *Creating the Kingdom of Ends*. Cambridge: Cambridge University Press.

Korsgaard, C. (1996b). *The Sources of Normativity*. Cambridge: Cambridge University Press.

Krimsky, S. and Golding, D. (1992). *Social Theories of Risk*. Westport: Praeger Publishers.

Lacewing, M. (2005). Emotional self-awareness and ethical deliberation. *Ratio*, 18: 65–81.

Lampugnani, V. (2006). The city of tolerant normality. In A. Graafland and L.J. Kavanaugh (eds.), *Crossover: Architecture, Urbanism & Technology*. Rotterdam: nai010 Publishers, pp. 294–312.

Landy, J.F. and Goodwin, G.P. (2015). Does incidental disgust amplify moral judgment? A meta-analytic review of experimental evidence. *Perspectives on Psychological Science*, 10: 518–536.

Lanting, C.I. and van Wouwe, J.P. (2007). *Redenen en motieven om te starten en te stoppen met borstvoeding*. Leiden: TNO-KvL, 2007. Publ. nr. 2007.105.

Lazarus, R. (1991). *Emotion and Adaptation*. New York: Oxford University Press.

Leiserowitz, A. (2005). American risk perceptions: Is climate change dangerous? *Risk Anal*, 25: 1433–1442.

Leiserowitz, A. (2006). Climate change risk perception and policy preferences: The role of affect, imagery, and values. *Climate Change*, 77: 45–72.

Levinson, J. (ed.) (1998). *Aesthetics and Ethics: Essays at the Intersection*. Cambridge: Cambridge University Press.

Little, M. and Halpern, J. (2009). Motivating health: Empathy and the normative activity of coping. In H. Lindemann, M. Walker, and M. Verkerk (eds.), *Naturalized Bioethics*. Cambridge: Cambridge University Press, pp. 141–162.

Little, M.O. (1995). Seeing and caring: The role of affect in feminist moral epistemology. *Hypatia: A Journal of Feminist Philosophy*, 10(3): 117–137.

Loewenstein, G. (1987). Anticipation and the valuation of delayed consumption. *The Economic Journal*, 97: 666–684.

Loewenstein, G. and Lerner, J.S. (2003). The role of affect in decision making. In R.J. Davidson, K.R. Scherer and H. Hill Goldsmith (eds.), *Handbook of Affective Sciences*. Oxford and New York: Oxford University Press, pp. 619–642.

Loewenstein, G.F., Weber, E.U., Hsee, C.K., and Welch, N. (2001). Risk as feelings. *Psychological Bulletin*, 127: 267–286.

Lorenzoni, I., Nicholson-Cole, S., and Whitmarsh, L. (2007). Barriers perceived to engaging with climate change among the UK public and their policy implications. *Global Environmental Change*, 17: 445–459.

Lorenzoni, I. and Pidgeon, N.F. (2006). Public views on climate change: European and USA perspectives. *Climatic Change*, 77: 73–95.

Lurie, Y. (2004). Humanizing business through emotions: On the role of emotions in ethics. *Journal of Business Ethics*, 49: 1–11.

Macilwain, C. (2011). Concerns over nuclear energy are legitimate. *Nature*, 471: 549.

Mackie, J.L. (1977). *Ethics: Inventing Right and Wrong*. Hammondsworth: Penguin.

Mampuys, R. and Roeser, S. (2011). Risk considerations in using GMO viruses as medicine; A conflict of emotions? *Journal of Disaster Research*, 6: 514–521.

Manders-Huits, N. and Zimmer, M. (2009). Values and pragmatic action: The challenges of introducing ethical intelligence in technical design communities. *International Review of Information Ethics*, 10: 37–44.

Marcus, G.E. (2000). Emotions in politics. *Annual Review of Political Science*, 3: 221–250.

Marcus, G.E. (2010). *The Sentimental Citizen: Emotion in Democratic Politics*. University Park: Penn State University Press.

Mayo, D.G. and Hollander, R.D. (eds.) (1994). *Acceptable Evidence: Science and Values in Risk Management*. Oxford: Oxford University Press.

McAllister, J.W. (2005). Emotion, rationality, and decision making in science. In H. Petr, V.-V. Luis, and W. Dag (eds.), *Logic, Methodology and Philosophy of Science: Proceedings of the Twelfth International Congress*. London: King's College Publications, pp. 559–576.

McDaniels, T.L., Gregory, R.S., and Fields, D. (1999). Democratizing Risk Management: Successful Public Involvement in Local Water Management Decisions. *Risk Analysis*, 19: 497–510.

McDowell, J. (1998). *Mind, Value and Reality*. Cambridge: Harvard University Press.

McNaughton, D. (1988). *Moral Vision*. Oxford: Basil Blackwell.

Meijnders, A.L., Midden, C.J.H., and Wilke, H.A.M. (2001). Role of negative emotion in communication about CO2 risks. *Risk Analysis*, 21: 955–966.

Mill, J.S., (1985 [1859]). *On Liberty*. Harmondsworth: Penguin.

Miller, W.I. (1997). *The Anatomy of Disgust*. Cambridge: Harvard University Press.

Möller, N. (2012). The concepts of risk and safety. In S. Roeser, R. Hillerbrand, P. Sandin, and M. Peterson (eds.), *Handbook of Risk Theory: Epistemology, Decision Theory, Ethics and Social Implications of Risk*. Dordrecht: Springer, pp. 55–85.

Möllering, G. (2001). The nature of trust: From Georg Simmel to a theory of expectation, interpretation and suspension. *Sociology*, 35: 403–420.

Moore, G.E. (1988 [1903]). *Principia Ethica*. Buffalo: Promotheus Books.

Moors, A. (2014). Examining the mapping problem in dual-process models. In J.W. Sherman, B. Gawronski, and Y. Trope (eds.), *Dual-Process Theories of the Social Mind*, New York: Guilford Press, pp. 20–34.

Moser, P.K. and Carson, T.L. (eds.) (2001). *Moral Relativism: A Reader*. New York and Oxford: Oxford University Press.

Moser, S.C. (2010). Communicating climate change: History, challenges, process and future directions. *WIRE's Climate Change*, 1: 31–53.

Mumby, D.K. and Putnam, L.L. (1992). The politics of emotion: A feminist reading of bounded rationality. *Academy of Management Review*, 17(3): 465–486.

Munster, A. (2013). *An Aesthesia of Networks: Conjunctive Experience in Art and Technology*. Cambridge: MIT Press.

Nasar, J.L. (ed.) (1988). *Environmental Aesthetics: Theory, Research, and Applications*. Cambridge: Cambridge University Press.

Neuman, R., Marcus, G., Crigler, A., and MacKuen, M. (2007). *The Affect Effect: Dynamics of Emotion in Political Thinking and Behavior*. Chicago: University of Chicago Press.

Nichols, S. (2004). *Sentimental Rules*. Oxford: Oxford University Press.

Nihlén Fahlquist, J. (2016). Experience of non-breastfeeding mothers: Norms and ethically responsible risk communication. *Nursing Ethics*, 23: 231–241.

Nihlén Fahlquist, J. and Roeser, S. (2011). Ethical problems with information on infant feeding in developed countries. *Public Health Ethics*, 4: 192–202.

Nihlén Fahlquist, J. and Roeser, S. (2015). Nuclear energy, responsible risk communication and moral emotions: A three level framework. *Journal of Risk Research*, 18(3): 333–346.

Nussbaum, M. (1992). *Love's Knowledge. Essays on Philosophy and Literature.* Oxford: Oxford University Press.

Nussbaum, M.C. (2001). *Upheavals of Thought: The Intelligence of Emotions.* Cambridge: Cambridge University Press.

Nussbaum, M.C. (2013). *Political Emotions: Why Love Matters for Justice.* Cambridge: Harvard University Press.

O'Neill, J. (2002). The rhetoric of deliberation: Some problems in Kantian theories of deliberative democracy. *Res Publica*, 8(3): 249–268.

Parfit, D. (1984). *Reasons and Persons.* Oxford: Oxford University Press.

Parfit, D. (2011). *On What Matters.* Oxford: Oxford University Press.

Pelser, A. (2010). Belief in Reid's theory of perception. *History of Philosophy Quarterly*, 27(4): 359–378.

Peterson, M. (2009). *An Introduction to Decision Theory.* Cambridge: Cambridge University Press.

Peterson, M. (2012). *The Dimensions of Consequentialism: Ethics, Equality and Risk.* Cambridge: Cambridge University Press.

Pinsky, M. (2003). *Future Present: Ethics and/as Science Fiction.* London: Associated University Press.

Plessner, H. (1928). *Die Stufen des Organischen und der Mensch. Einleitung in die philosophische Anthropologie.* Berlin: de Gruyter.

Prichard, H.A. (1912). Does moral philosophy rest on a mistake? *Mind*, 21: 21–37.

Rachels, J. (1999). *The Elements of Moral Philosophy.* New York: McGraw-Hill.

Rawls, J. (1971). *A Theory of Justice.* Cambridge: Harvard University Press.

Rawls, J. (1996). *Political Liberalism.* New York: Columbia University Press.

Reichle, I. (2009). *Art in the Age of Technoscience: Genetic Engineering, Robotics, and Artificial Life in Contemporary Art.* Vienna and New York: Springer.

Reid, T. (1969a [1785]). *Essays on the Intellectual Powers of Man.* Introduction by Baruch Brody. Cambridge and London: The MIT Press.

Reid, T. (1969b [1788]). *Essays on the Active Powers of the Human Mind.* Introduction by Baruch Brody. Cambridge and London: The MIT Press.

Renn, O. (1998). The role of risk perception for risk management. *Reliability Engineering & System Safety*, 59: 49–62.

Roberts, R.C. (2003). *Emotions: An Essay in Aid of Moral Psychology.* Cambridge and New York: Cambridge University Press.

Robinson, J.G. and Mcllwee, J.S. (2005). Men, women, and the culture of engineering. *Sociological Quarterly*, 32(3): 403–421.

Roeser, S. (2005). Intuitionism, moral truth, and tolerance. *Journal of Value Inquiry*, 39: 75–87.

Roeser, S. (2006). The role of emotions in judging the moral acceptability of risks. *Safety Science*, 44(8): 689–700.

Roeser, S. (2007). Ethical intuitions about risks. *Safety Science Monitor*, 11: 1–30.

Roeser, S. (2009), 'Reid and Moral Emotions', *Journal of Scottish Philosophy,* 7: 177–192.

Roeser, S. (2010). Intuitions, emotions and gut feelings in decisions about risks: Towards a different interpretation of 'neuroethics'. *The Journal of Risk Research*, 13: 175–190.

Roeser, S. (2011a). *Moral Emotions and Intuitions.* Basingstoke: Palgrave Macmillan.

Roeser, S. (2011b). Nuclear energy, risk and emotions. *Philosophy and Technology*, 24: 197–201.

Roeser, S. (2012). Risk communication, public engagement, and climate change: A role for emotions. *Risk Analysis*, 32: 1033–1040.

Roeser, S., Alfano, V., and Nevejan, C. (2017). Risk, art and moral emotions. *Ethical Theory and Moral Practice* (forthcoming).

Roeser, S., Hillerbrand, R., Peterson, M., and Sandin, P. (eds.) (2012). *Handbook of Risk Theory: Epistemology, Decision Theory, Ethics, and Social Implications of Risk*. Dordrecht: Springer.

Roeser, S. and Nihlén Fahlquist, J. (2014). Moral emotions and risk communication. In J. Arvai and L. Rivers (eds.), *Effective Risk Communication*. London: Earthscan/ Routledge, pp. 204–219.

Roeser, S. and Pesch, U. (2016). An emotional deliberation approach to risk. *Science, Technology & Human Values*, 41: 274–297.

Roeser, S. and Todd, C. (eds.) (2014). *Emotion and Value*. Oxford: Oxford University Press.

Rorty, R. (1989). *Irony, Contingency, and Solidarity*. Cambridge: Cambridge University Press.

Roser-Renouf, C. and Maibach, E. (2010). Communicating climate change. In S. Hornig Priest (ed.), *The Encyclopedia of Science and Technology Communication*. London: Sage.

Ross, A. and Athanassoulis, N. (2012). Risk and virtue ethics. In S. Roeser, R. Hillerbrand, P. Sandin, and M. Peterson (eds.), *Handbook of Risk Theory: Epistemology, Decision Theory, Ethics and Social Implications of Risk*. Dordrecht: Springer, pp. 833–856.

Ross, W.D. (1927). The basis of objective judgments in ethics. *International Journal of Ethics*, 37: 113–127.

Ross, W.D. (1967 [1930]). *The Right and the Good*. Oxford: The Clarendon Press.

Ross, W.D. (1968 [1939]). *Foundations of Ethics: The Gifford Lectures*. Oxford: Clarendon Press.

Rothman, A.J. and Salovey, P. (1997). Shaping perceptions to motivate healthy behavior: The role of message framing. *Psychological Bulletin*, 121: 3–19.

Samuelson, W. and Zeckhauser, R. (1988). Status quo bias in decision making. *Journal of Risk and Uncertainty*, 1(1): 7–59.

Sandman, P.M. (1989). Hazard versus outrage in the public perception of risk. In V.T. Covello, D.B. McCallum, and M.T. Pavlova (eds.), *Effective Risk Communication: The Role and Responsibility of Government and Nongovernment Organizations*. New York: Plenum Press, pp. 45–49.

Sauer, H. (2014). The wrong kind of mistake: A problem for robust sentimentalism about moral judgment. *The Journal of Value Inquiry*, 48: 247–269.

Scheler, M. (1948). *Wesen und Formen der Sympathie*. Frankfurt/Main: Schulte-Bulenke.

Scherer, K.R. (1984). On the nature and function of emotion: A component process approach. In K.R. Scherer and P. Ekman (eds.), *Approaches to Emotion*. Hillsdale and London: Lawrence Erlbaum Associates, pp. 293–317.

Scheufele, D.A., Corley, E.A., Dunwoody, S., Shih, T.-J., Hillback, E., and Guston, D.H. (2007). Scientists worry about some risks more than the public. *Nature Nanotechnology*, 2: 732–734.

Schivelbusch, W. (1986 [1977]). *The Railway Journey: The Industrialization of Time and Space in the 19th Century*. Berkeley: University of California Press.

Schwarz, N. (2002). Feelings as information: Moods influence judgments and processing strategies. In T. Gilovich, D. Griffin, and D. Kahnemann (eds.),

Intuitive Judgment: Heuristics and Biases. Cambridge: Cambridge University Press, pp. 534–547.

Schwitzgebel, E. (2010). Acting contrary to our professed beliefs, or The Gulf between occurrent judgment and dispositional belief. *Pacific Philosophical Quarterly*, 91: 531–553.

Sclove, R.E. (2000). Town Meetings on Technology: Consensus Conferences as Democratic Participation. In Daniel Lee Kleinman (ed.), *Science, Technology, and Democracy*. New York: State University of New York Press

Scruton, R. (1979). *The Aesthetics of Architecture*. Chetham: Mackay Limited.

Shafer-Landau, R. (2003). *Moral Realism: A Defence*. Oxford: Oxford University Press.

Sheppard, S.R.J. (2005). Landscape visualisation and climate change: The potential for influencing perceptions and behavior. *Environmental Science & Policy*, 8: 637–654.

Sherman, N. (1989). *The Fabric of Character: Aristotle's Theory of Virtue*. Oxford: Clarendon Press.

Shrader-Frechette, K.S. (1991). *Risk and Rationality: Philosophical Foundations for Populist Reforms*. Berkeley: University of California Press.

Sidgwick, H. (1901 [1874]). *The Methods of Ethics*. London and New York: Macmillan.

Simon, H. (1987). Making management decisions: The role of intuition and emotion. *The Academy of Management Executive*, 1(1): 57–64.

Singer, P. (2005). Ethics and intuitions. *The Journal of Ethics*, 9: 331–352.

Sinnot-Armstrong, W. (2006). Moral intuitionism meets empirical psychology. In T. Horgan and M. Timmons (eds.), *Metaethics After Moore*. Oxford: Clarendon Press.

Sjöberg, L. (2006). Will the real meaning of affect please stand up? *Journal of Risk Research*, 9: 101–108.

Sloman, S.A. (1996). The empirical case for two systems of reasoning. *Psychological Bulletin*, 119(1): 3–22.

Sloman, S.A. (2002). Two systems of reasoning. In T. Gilovich, D.W. Griffin, and D. Kahneman (eds.), *Heuristics and Biases: The Psychology of Intuitive Judgment*. Cambridge: Cambridge University, pp. 379–396.

Slovic, P. (1992). Perception of risk: Reflections on the psychometric paradigm. In S. Krimsky and D. Golding (eds.), *Social Theories of Risk*. New York: Praeger, pp. 117–152.

Slovic, P. (1999). Trust, emotion, sex, politics, and science: Surveying the risk-assessment battlefield. *Risk Analysis*, 19: 689–701.

Slovic, P. (2000). *The perception of risk*. London: Earthscan.

Slovic, P. (2007). Numbed by numbers. *Foreign Policy*. (March 13, 2007), Available at: http://foreignpolicy.com/2007/03/13/numbed-by-numbers/.

Slovic, P. (2010a). *The Feeling of Risk: New Perspectives on Risk Perception*. London: Earthscan.

Slovic, P. (2010b). 'If I look at the mass I will never act': Psychic numbing and genocide. In S. Roeser (ed.), *Emotions and Risky Technologies*. Springer: Dordrecht, pp. 37–59.

Slovic, P., Finucane, M., Peters, E., and MacGregor, D.G. (2002). The affect heuristic. In T. Gilovich, D. Griffin, and D. Kahnemann (eds.), *Heuristics and Biases: The Psychology of Intuitive Judgement*. Cambridge: Cambridge University Press, pp. 397–420.

Slovic, P., Finucane, M., Peters, E., and MacGregor, D.G. (2004). Risk as analysis and risk as feelings: Some thoughts about affect, reason, risk, and rationality. *Risk Analysis*, 24(2): 311–322.

Slovic, S. and Slovic, P. (eds.) (2015). *Numbers and Nerves: Information, Emotion, and Meaning in a World of Data*. Corvallis: Oregon State University Press.

Solomon, R.C. (1993). *The Passions: Emotions and the Meaning of Life*. Indianapolis: Hackett Publishing Company.

Spence, A. and Pidgeon, N.F. (2010). Framing and communicating climate change: The effects of distance and outcome frame manipulations. *Global Environmental Change*, 20: 656–667.

Staiger, J., Cvetkovich, A., and Reynolds, A. (eds.) (2010). *Political Emotions*. New York: Routledge.

Stanovich, K.E. and West, R.F. (2002). Individual differences in reasoning: Implications for the rationality debate? In T. Gilovich, D.W. Griffin, and D. Kahneman (Eds.), *Heuristics and Biases: The Psychology of Intuitive Judgment*. Cambridge: Cambridge University Press, pp. 421–440.

Stirling, A. (2002). Multi-Criteria Mapping: Mitigating the Problems of Environmental Valuation? In J. Foster (ed.), *Valuing Nature?: Economics, Ethics and Environment*. London and New York: Routledge, pp. 186–210.

Stocker, M. with Hegemann, E. (1996). *Valuing Emotions*. Cambridge: Cambridge University Press.

Stratton-Lake, P. (ed.) (2002). *Ethical Intuitionism: Re-evaluations*. Oxford: Oxford University Press.

Sunderland, M. (2013). Taking emotion seriously: Meeting students where they are. *Science and Engineering Ethics*, 20: 183–195.

Sunstein, C.R. (2002). *Risk and Reason: Safety, Law, and the Environment*. Cambridge: Cambridge University Press.

Sunstein, C.R. (2005). *Laws of Fear*. Cambridge: Cambridge University Press.

Szabó Gendler, T. (2008). Alief and belief. *Journal of Philosophy*, 105(10): 634–663.

Taebi, B. (2012a). Intergenerational risks of nuclear energy. In S. Roeser, R. Hillerbrand, P. Sandin, and M. Peterson (eds.), *Handbook of Risk Theory: Epistemology, Decision Theory, Ethics and Social Implications of Risk*. Dordrecht: Springer, pp. 295–318.

Taebi, B. (2012b). Multinational nuclear waste repositories and their complex issues of justice. *Ethics, Policy & Environment*, 15: 57–62.

Taebi, B. and Roeser, S. (eds.) (2015). *The Ethics of Nuclear Energy*. Cambridge: Cambridge University Press.

Taebi, B., Roeser, S., and van de Poel, I. (2012). The ethics of nuclear power: Social experiments, intergenerational justice, and emotions. *Energy Policy*, 51: 202–206.

Taylor, N. (2000). Ethical arguments about the aesthetics of architecture. In W. Fox (ed.), *Ethics and the Built Environment*. London: Routledge, pp. 193–206.

Thaler, R.H. and Sunstein, C.R. (2008). *Nudge: Improving Decisions About Health, Wealth and Happiness*. New York: Penguin.

Tropman, E. (2009). Renewing moral intuitionism. *Journal of Moral Philosophy*, 6(4): 440–463.

Tversky, A. and Kahneman, D. (1974). Judgment under uncertainty: Heuristics and biases. *Science*, 185: 1124–1131.

van Asselt, M. and Rijkens-Klomp, N. (2002). A look in the mirror: Reflection on participation in integrated assessment from a methodological perspective. *Global Environmental Change*, 12: 167–184.

van Asselt, M. and Vos, E. (2006). The precautionary principle and the uncertainty paradox. *Journal of Risk Research*, 9: 313–336.

van den Hoven, J. (2007). ICT and value sensitive design. In P. Goujon, S. Lavelle, P. Duquenoy, K. Kimppa, and V. Laurent (eds.), *The Information Society: Innovation, Legitimacy, Ethics and Democracy*. Boston: Springer.

van den Hoven, J. (2014). Responsible innovation: A new look at technology and ethics. In J. van den Hoven, N. Doorn, T. Siwerstra, B.-J. Koops, and H. Romijn (eds.), *Responsible Innovation*. Dordrecht: Springer, pp. 3–13.

van de Poel, I. (2012). Can we design for well-being? In P. Brey, A. Briggle, and E. Spence (eds.), *Good Life in a Technological Age*. New York: Routledge, pp. 295–306.

van de Poel, I. (2013). Why new technologies should be conceived as social experiments. *Ethics, Policy & Environment*, 16: 352–355.

van de Poel, I. and Nihlén Fahlquist, J. (2012). Risk and responsibility. In S. Roeser, R. Hillerbrand, M. Peterson, and P. Sandin (eds.), *Handbook of Risk Theory: Epistemology, Decision Theory, Ethics, and Social Implications of Risk*. Dordrecht: Springer, pp. 877–907.

van der Burg, S. and van Gorp, A. (2005). Understanding moral responsibility in the design of trailers. *Science and Engineering Ethics*, 11: 235–256.

Veen, M., te Molder, H., Gremmen, B., and van Woerkum, C. (2010). Quitting is not an option: An analysis of online diet talk between celiac disease patients. *Health: An Interdisciplinary Journal for the Study of Health, Illness and Medicine*, 14: 23–40.

Venturi, R., Brown, D.S., and Izenour, S. (1972). *Learning From Las Vegas*. Cambridge: MIT Press.

Weber, E.U. (2006). Experience-based and description-based perceptions of long-term risk: Why global warming does not scare us (yet). *Climatic Change*, 77: 103–120.

Weichman, J.C. (ed.) (2008). *The Aesthetics of Risk*. Zurich: JRP|Ringier Books.

Weiner, R.F. (1993). Comment on Sheila Jasanovs guest editorial. *Risk Analysis*, 13: 495–496.

Wellman, C. (1963). The ethical implications of cultural relativity. *Journal of Philosophy*, 60(7): 169–184.

Whewell, W. (1845). *The Elements of Morality, Including Polity*. London: John W. Parker.

Williams, B. (1973). A critique of utilitarianism. In J.J.C. Smart and B. Williams (eds.), *Utilitarianism: For and Against*. Cambridge: Cambridge University Press.

Willis, H.H. and DeKay, M.L. (2007). The roles of group membership, beliefs, and norms in ecological risk perception. *Risk Analysis*, 27: 1365–1380.

Willis, H.H., DeKay, M.L., Fischhoff, B., and Morgan, G. (2005). Aggregate, disaggregate, and hybrid analyses of ecological risk perceptions. *Risk Analysis*, 25: 405–428.

Wilson, S. (2002). *Information Arts. Intersections of Art, Science, and Technology*. Cambridge: MIT Press.

Wolf, J.B. (2011). *Is Breast Best? Taking on the Breastfeeding Experts and the New High Stakes of Motherhood*. New York and London: New York University Press.

Wolff, J. (2006). Risk. *The Encyclopedia of Political Thought*, 3281–3283. DOI: 10.1002/9781118474396.wbept0895

Wollheim, R. (1999). *On the Emotions*. New Haven and London: Yale University Press.

Zagzebski, L. (2003). Emotion and moral judgment. *Philosophy and Phenomenological Research*, 66: 104–124.

Zajonc, R.B. (1980). Feeling and thinking: Preferences need no inferences. *American Psychologist*, 35: 151–175.

Zajonc, R.B. (1984a). The interaction of affect and cognition. In K.R. Scherer and P. Ekman (eds.), *Approaches to Emotion*. Hillsdale and London: Lawrence Erlbaum Associates, pp. 239–257.

Zajonc, R.B. (1984b). On primacy of affect. Approaches to emotion. In K.R. Scherer and P. Ekman (eds.), *Approaches to Emotion*. Hillsdale and London: Lawrence Erlbaum Associates, pp. 259–270.

Zangwill, N. (2003). Against moral response-dependence. *Erkenntnis*, 59: 285–290.

Zinn, J.O. (2008). Heading into the unknown: Everyday strategies for managing risk and uncertainty. *Health, Risk & Society*, 10: 439–450.

Zwijnenberg, R. (2009). Preface. In I. Reichle (ed.), *Art in the Age of Technoscience: Genetic Engineering, Robotics, and Artificial Life in Contemporary Art*. Vienna and New York: Springer, pp. xiii–xxix.

Zwijnenberg, R. (2014). Biotechnology, human dignity and the importance of art. *Teoria: Revista di Filosofia*, 131–148.

Index

Note: Page references in *italics* denote tables.